Make: High-Power Rockets

Construction and Certification for Thousands of Feet and Beyond

Mike Westerfield

MAKER MEDIA™
SAN FRANCISCO, CA

Make: High-Power Rockets

by Mike Westerfield

Published by Maker Media, Inc., 1160 Battery Street East, Suite 125, San Francisco, CA 94111.

Maker Media books may be purchased for educational, business, or sales promotional use. Online editions are also available for most titles (*http://oreilly.com/safari*). For more information, contact O'Reilly Media's institutional sales department: 800-998-9938 or *corporate@oreilly.com*.

Editor: Patrick Di Justo	**Indexer:** WordCo Indexing Services
Production Editor: Nicholas Adams	**Interior Designer:** David Futato
Copyeditor: Jasmine Kwityn	**Cover Designer:** Julie Cohen
Proofreader: Rachel Head	**Illustrator:** Rebecca Demarest

November 2017:   First Edition

Revision History for the First Edition
2017-11-01:   First Release

See *http://oreilly.com/catalog/errata.csp?isbn=9781457182976* for release details.

978-1-4571-8297-6

[TI]

FRESH FISH

FRESH FISH

A FEARLESS GUIDE TO
GRILLING, SHUCKING, SEARING, POACHING
AND ROASTING SEAFOOD

JENNIFER TRAINER THOMPSON

PHOTOGRAPHS BY KELLER + KELLER

*The mission of Storey Publishing is to serve our customers by
publishing practical information that encourages
personal independence in harmony with the environment.*

EDITED BY Margaret Sutherland and Lisa Hiley
ART DIRECTION AND BOOK DESIGN BY Carolyn Eckert
TEXT PRODUCTION BY Jennifer Jepson Smith
INDEXED BY Christine R. Lindemer, Boston Road Communications

COVER AND INTERIOR PHOTOGRAPHY BY © Keller + Keller Photography, except as noted below
PROP AND FOOD STYLING BY Catrine Kelty

ADDITIONAL PHOTOGRAPHY BY
Carolyn Eckert, 15 (left), 24, 165, 175 (bottom), 213, 237 (bottom), 240, 257, 281, 282;
courtesy of the author, 11 (bottom right), 12 (top), 170 (top);
© Emma Kim/Getty Images, 179;
© Jennifer Trainer Thompson, 86 (bottom), 157;
© Julie Bidwell, 219;
© Katie Craig, 79, 86 (top left & right), 122, 133, 138, 142, 207, 262;
Mars Vilaubi, 59 (bottom), 127, 192, 337;
© Stacey Cramp, 141; © Wayne Smith, 51, 85, 89, 95, 98 (top left)
Nautical charts from the National Oceanic and Atmospheric Administration (NOAA),
U.S. Department of Commerce

Storey Publishing
210 MASS MoCA Way
North Adams, MA 01247
www.storey.com

Printed in China by R.R. Donnelley
10 9 8 7 6 5 4 3 2 1

Library of Congress Cataloging-in-Publication Data
is on file

"I look for places that exist out of time.
I look for an inarticulable quality
to architecture that bears the traces of
human history. That, combined
with the people who inhabit a place,
its natural elements,
all add up to its mood, its aura,
its visual character."

— ARTIST GREGORY CREWDSON

CONTENTS

INTRODUCTION

LIKE MY PARENTS, I grew up along the New England coast. I sailed before I could walk, and thought nothing of the fact that chapters in my elementary school history books — tales of witch hunts, Paul Revere's ride, John and Abigail — were about characters from nearby towns. Although we moved a lot when I was a child, when I reflect on the patchwork quilt of my earliest memories — catching an eel off the dock in Hingham, sailing across Casco Bay on Saturday mornings to get doughnuts at Handy's Boat Yard, swimming with my great-aunts at the Fairfield Beach Club — it is always with an eye toward the sea.

In 1963, we moved to Indianapolis (Indian-no-place, my father called it), and the summer heat coming off the soybean fields across the street hit my face like a hot furnace blast. While my father built a Sunfish in the garage, I made toy boats that I tried to sail in the tornado ditches after heavy rains. I was the only kid in the neighborhood whose parents called her home to dinner with a foghorn. When I was sent to a camp where the idea of waterfront activity was catching crawdads in a buggy creek, I lasted a week (I wasn't much for playing softball in the heat) but was perfectly content to play in my neighborhood instead. On the day we moved in, I had knocked on the neighbor's door to see if they had any kids my age and had stumbled upon Julie Weintraub's seventh birthday party. She not only became my best friend, but at her party I met all the kids in the neighborhood, a 1960s development of brick ranch houses, square unfenced yards, and Hoosier-neighbor friendliness.

My mother imported Maine lobsters and threw a clambake for the neighbors, who'd never seen a lobster. She good-naturedly took their ribbing when she talked about *pahked cahs in Hahvad Yahd*, volunteered in the inner city to help get Head Start off the ground, and rooted for Bobby Kennedy in a sea of Republicans.

The next spring she and my father, worried that I might be acquiring a Hoosier accent and forgetting the glories of New England summers, suggested that I spend the season with relatives on Buzzards Bay in Massachusetts. I thought it was a swell idea.

Imagine, at age eight boarding a plane from Indianapolis to Boston without an escort. On the tarmac (in those days you walked outside to board a commercial plane) I got a little teary, and I'll never forget my mother stopping, looking me in the eye, and saying, "Jenny, you don't have to go if you don't want to."

Those liberating words propelled me to adventures that became part of my DNA.

SUMMERS ON BUZZARDS BAY

FOR THE NEXT FOUR YEARS, I summered (we never called it that, but it's sometimes a good thing when a noun morphs into a verb) with cousins in Wareham. In a small cottage at the confluence of the WeWeantic and Sippican rivers (my father joked we should start a Sippi-WeWe Yacht Club), our days were spent barefoot in bathing suits — we'd switch into a dry one as we hung a wet one on a nail in the outdoor shower. High tides were for swimming, sailing, and water-skiing when the weather was right. Low tides were spent playing cards (hours upon hours of canasta), doing jigsaw puzzles, clamming, and bike riding. Early mornings were for fishing (I could go with my uncles and male cousins only if I unhooked and cleaned my own fish). We had no TV. My aunt was our pretty, blonde camp director who organized hootenannies every Sunday evening (this was the late '60s, after all) and would chauffeur us in her powder-blue Mustang on nights we wanted to go mini-golfing or go-karting and get soft ice cream. With plenty of relatives around, a big trip might be to Nauset Beach, on the outer shore of Cape Cod ("the Cape"), where we'd try our luck surfing, stopping to dunk in a kettle pond to rinse off the salt on our way home. Once a summer we'd board the ferry to Martha's Vineyard for a day of bike riding, cold surf swimming, and saltwater taffy.

Sure, there exists a Brahmin New England, where obscure references serve as shorthand for those acquainted with privilege and wealth. If you're traveling to

"Someone, please reel her in. Oh, Cod."

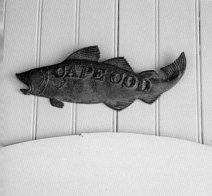

This is a cookbook, yes, but it's also a barefoot ode to the soul of the coast.

Pulpit Harbor . . . Kittansett . . . Naushon . . . these simple words speak volumes if you're in the know. But there are also the Swamp Yankees, as my dad called us, descended from the original Puritans, who may be low on cash but rich on heritage and know these waters because for centuries their forebears depended on them for sustenance — and their descendants still know when the stripers and blues are running (September), bay scallops are harvested (November), and beach plums are ripe (late August). Cranberry bogs serve two purposes: providing a star dish at the nation's table from November 1 to December 26, and a perfect outdoor skating rink.

SAILING DOWN THE COAST

AFTER COLLEGE, I SPENT A SUMMER waitressing on Maine's Mount Desert Island, where my cousin suggested we sign on for a sailboat delivery to the Caribbean on a guy's Bermuda 40. (This is a bit like being asked to drive an Alfa Romeo down the Amalfi Coast.) We had a crew of four: the captain, first mate (my cousin), second mate (me), and cook (the captain was no fool). The first leg — from northern Maine to southern Connecticut — covered the territory in this book. We dug clams before we left Southwest Harbor and kept a pot of chowder on the gimbaled stove as we made it down Massachusetts Bay, through the Cape Cod Canal, into Buzzards Bay, Narragansett Bay, then past Fishers Island and Stamford to Hell's Gate.

Some of the coast has changed since then — it's (thankfully) hard to sail past a rocky point without finding farmers' markets or locally sourced cheeses, vegetables, herbs, or seafood. The shoreline is also more populated, as unheated summer cottages are converted to wireless year-round homes, and improved transportation expands the commuting range from Portland, Boston, Providence, and New Haven. Sure, environmental causes (climate change, ocean warming, acidification, invasives like green crabs) are affecting certain traditional New England fisheries, such as the decline Maine is seeing in northern shrimp, lobster, smelt, and soft-shell clams. But thanks to conscientious environmental stewardship, despite overfishing, the waters and harbors are also cleaner than they've been in years, and oysters and other shellfish have returned in abundance.

And some things remain the same — the tides flow in twice daily, bringing stripers and blues. Clamming and quahogging techniques haven't changed since Native Americans taught the Pilgrims. New England flavors are uniquely American and regional, characterized by an abundance of seafood (it's amazing that the Pilgrims almost starved that first winter, given the wealth of cod in the waters), and many ethnic influences, from the Puritans to the thousands of immigrants who arrived as crew on whaling ships and later to work in the mills. Great seafood — the heart of coastal New England cooking, due to the cold waters — deserves top billing. If it's impeccably fresh, the simplest treatment will shine.

THE TASTE OF A REGION

SO WHY THIS BOOK, AND WHY NOW? Plenty of great food traditions are perfectly intact, and have been for hundreds of years, and a few of those are included here, such as clambakes on the beach. But coastal New England today is far more interesting from a culinary point of view than it used to be, what with aquaculture, artisanal cheeses, spirits, beer, seafood, organic farms, and oysters. Thank goodness Boston baked beans are no longer the only side dish — though even those are being jazzed up. There's a reason why so many great chefs — Chris Schlesinger, Johanne Killeen, the late George German — have flocked to the Atlantic seaboard. In small New England harbors, cities, peninsulas, and islands, there's a playful pursuit of excellence, a sense of community, and a dedication to authenticity (whatever that might be), whether it's roasting, brewing, harvesting, coddling, digging, picking, churning, or throwing out a line. In many places, I found an invigorating lack of attitude and disregard for fads. Regional flavors are being modernized without losing the character and vitality this region has always drawn from the North Atlantic. While nodding still to the old New England traditions, the food is better — seasonal, contemporary, exuberant, eccentric.

And it's not hard. If you know my books, you know my motto is fresh flavors, nicely tuned, with minimal fuss. Yet I'm surprised by the number of people who make excuses for not cooking fish — it will smell up the kitchen (not necessarily true), it sticks to the grill (doesn't have to), it's not interesting (a Catholic hangover from fish stick Fridays) — but really I think many people don't do it because they are

intimidated. Fish *is* sort of weird looking (never mind the bivalves). It's slippery. And those eyes: you don't see eyes staring at you when you buy or are served cow. Let's get over it. Fish and shellfish are superfoods — lean, one of the world's most protein-rich foods, good for your brain and heart. They are remarkably easy to roast in the winter, pan-fry in the fall, stir-fry in the spring, and grill year-round. They are the original 20-minute gourmet. Go fish!

Years ago, when I was waitressing at a diner in Northeast Harbor, Maine, an old gentleman came in every day for a slice of pie. A real sweetie, he had retired as a chauffeur at one of the great estates, and boasted that he owned one suit, a seersucker pinstripe from Brooks Brothers that the owner of the estate had bought him years ago to attend a funeral. He sat on a stool at the end of the counter, waiting for his daily slice, when the umpteenth tourist came in asking the way to Bar Harbor.

"Just follow the yellow line," he replied drily.

Please follow the yellow line with me down the coast, from Connecticut to Maine — savoring the stories, history, seafood, recipes, and exuberance of coastal New England. This is a cookbook, yes, but it's also a barefoot ode to the soul of the coast and the profound pleasures of being by the water. As my father and uncles would say when someone got the ball rolling after supper at the cottage: it may be a little fishy — but don't clam up — or carp about it — you gotta have mussel — just for the halibut — maybe that's allureing? — don't flounder — come out of your shell — you'll get hooked. Someone, please reel her in. Oh, Cod.

— JENNIFER TRAINER THOMPSON, MOLLY'S COVE

SOUPS AND CHOWDERS

W HAT A GREAT WORD: CHOWDER. "Chowdah," my mother always said, flipping it off the roof of her mouth with gusto. A hodgepodge of flavors and textures, chowder was probably invented on the coasts of northwestern France and southwestern England in the 16th century, a make-do meal for fishermen that combined the day's catch with what was in the garden or on shipboard. The word's origin is thought perhaps to be *chaudière*, French for large iron pot, but another possible root is *jowter*, Cornish for fishmonger. When ships returned home from months at sea, the town's welcome celebration included a communal chaudière, into which each fisherman contributed part of his haul.

Fishing as far out as the Grand Banks, early explorers probably took their catch ashore in Newfoundland, and a stew was a smart way to cook shellfish and fish with shipboard provisions. Early European settlers found that Native Americans were already making chowders with shellfish. Indeed, in some areas clams and oysters were eaten in such prodigious quantity that you can still find oyster shell mounds piled 10 feet high — there's one in my neighborhood in Mattapoisett, hundreds of years old, from when the Wampanoags summered there. The Pilgrims, who turned their noses up at shellfish, fed clams and mussels to their hogs, calling shellfish "the meanest of God's blessings."

The first printed chowder recipe in Massachusetts was in the *Boston Evening Post* in 1751 (seasoned with salt, pepper, marjoram, savory, thyme, and parsley), and chowder remains a New England staple. Interestingly, the 1896 *Boston Cooking School Cook Book* had only one recipe for clam chowder, and it was prepared with the traditional cream base. Later editions included Manhattan Chowder (water-based, with tomatoes) and Rhode Island Clam Chowder (clear broth, with bacon).

When my parents married in 1952, they spent their honeymoon camping on the beach in Provincetown, where they kept a pot of chowder continually on the fire, replenishing it with each day's catch. Though quahogs (large hard-shell clams) are the traditional chowder clam because they are large and tough (quahogs are never served raw), any type of clam and most kinds of fish can be used in a chowder, which improves with age. For those who haven't cooked much with fish, a one-pot dish is a great introduction. Rustic and flexible, chowders shine with flavor. They are also profoundly simple, liberating the cook for other pleasures.

Rustic and
flexible,
chowders shine
with flavor.

Clam Chowder

New England chowders are traditionally made with quahogs — large hard-shell clams that are native to the eastern shores of North America and particularly plentiful near Cape Cod and the Islands. Littleneck clams, which are smaller hard-shell clams, would also be tasty in this dish.

48	hard-shell (littleneck) clams, scrubbed (for 2 cups meat); see page 41
4–5	slices bacon, chopped
6	tablespoons butter
2	celery stalks, chopped
1	medium yellow onion, diced
½	teaspoon dried thyme or 1 teaspoon fresh
¼	teaspoon cayenne pepper
6	tablespoons all-purpose flour
3	large potatoes, diced
2	cups heavy cream
2	cups milk
	Butter, for serving
	Freshly ground black pepper

1. Bring 6 cups water to a boil in a soup pot and add the clams. Return to a boil, then reduce the heat to medium and steam until the clams open, about 7 minutes. Use a slotted spoon to transfer the clams to a bowl. Discard any clams that did not open. Strain the clam broth through a paper towel–lined colander, reserving 4 cups of the broth. Shell the clams and mince the meat.

2. In the soup pot, fry the bacon until crispy, about 8 minutes. Using a slotted spoon, transfer the bacon to a paper towel–lined plate and discard the fat in the pot. Melt the butter in the soup pot. Add the celery, onion, thyme, and cayenne, and cook over medium heat for 10 minutes, stirring occasionally. Add the flour and cook 2 to 3 minutes, stirring frequently. Add the reserved broth from the cooked clams and stir until thickened slightly, about 5 minutes. Add the potatoes and cream, reduce the heat to low, and cook until the potatoes are tender, about 20 minutes.

3. Add the milk, bring the chowder to a boil, and then reduce the heat to medium and add the reserved clams and bacon. Simmer until the clams are heated through, about 5 minutes. Serve in bowls topped with pats of butter if you wish and a generous grinding of black pepper.

SERVES 8

"CHOWDER BREATHES REASSURANCE.
IT STEAMS CONSOLATION."
—CLEMENTINE PADDLEFORD

Rhode Island Clam Chowder

Rhode Islanders prefer a clear-broth chowder to the traditional "white" chowder, as they deign to call chowders across the state line. If you've never tried a clear-broth chowder, rush thee to the kitchen: it's awesome. From Galilee to Warwick, Rhode Island Broth Chowder is dished up in schools, at diners, in pubs. Indeed, I learned the secret from a public school line cook who moonlights at an oyster bar in Jamestown: start by rendering fat from salt pork and then make a roux by adding flour to the fat. The flavors are strong and pronounced, and I think that a clam broth is better than chicken soup for a cold. I like it with a lot of black pepper and a few dashes of Tabasco sauce.

20	littleneck clams
4	slices bacon, chopped
2	tablespoons butter
1	large yellow onion, diced
2	celery stalks, diced
2	teaspoons fresh thyme
½	teaspoon black pepper
¼	teaspoon garlic powder
¼	teaspoon salt
3	tablespoons all-purpose flour
1	pound yellow potatoes, diced

1. Bring 6 cups water to a boil in a soup pot and add the clams. Return to a boil, then reduce the heat to medium and steam until the clams open, about 7 minutes. Use a slotted spoon to transfer the clams to a bowl. Discard any clams that did not open. Strain the clam broth through a paper towel–lined colander, reserving 4 cups of the broth. Shell the clams and dice the meat.

2. Fry the bacon in the soup pot over medium heat until crisp, about 8 minutes. With a slotted spoon, remove the bacon, leaving 2 tablespoons fat in the pot. Add the butter, onion, celery, thyme, pepper, garlic powder, and salt, and cook until the onions are translucent, about 7 minutes.

3. Add the flour, stir, and cook for 2 to 3 minutes. Add the reserved broth, stir, and then add the potatoes and the reserved bacon and cook until the potatoes are tender. Stir in the reserved clams to warm them, then turn off the heat.

SERVES 4-6

LITTLE RHODY

A small state with big flavor, Rhode Island is home to stuffies (stuffed quahogs; see page 167) and Del's Lemonade (page 275), not to mention clam cakes (page 162), a masterful street snack. The first colony to declare its independence and the last to ratify the Constitution, Little Rhody insists on calling a milkshake a cabinet, a sub a grinder, and their signature chowder a broth.

A mere 37 miles from east to west, the Ocean State has a 420-mile coastline of deep bays and low barrier beaches, with 33 islands in Narragansett Bay, an estuary that reaches two-thirds of the way up the state. The fact that no Rhode Islander is more than 30 minutes from the water informs their culinary sensibilities, to be sure.

In the early 1600s, about 4,000 Narragansetts and 1,500 Wampanoags lived in the area now called Rhode Island. Seafood was a staple; they caught striped bass with bone hooks and nets and gathered quahogs, oysters, and other shellfish from shallow waters.

In the 1630s, Puritan theologian Roger Williams fled persecution in Massachusetts and founded Providence. More European settlers arrived, and by the early 19th century, English, Irish, and Scottish settlers were arriving in droves, followed later by Portuguese, Italian, and Polish immigrants. They brought their food heritages with them when they came to work in the mills, creating a vibrant and diverse food culture in a small state that many motorists breeze through on their way from New York to the Martha's Vineyard. Meanwhile, the million residents of the Ocean State tuck in and stubbornly maintain their food heritage, tradition, terminology, and predilections in a small area.

24

JFK New England Fish Chowder

Thank goodness for secretaries. When John F. Kennedy was president, a disabled girl wrote to him asking what he liked to eat. "Please reply to her," Kennedy's secretary wrote in a memo to the president. "She will be extremely happy. Do not mention anything in the letter about her handicap *please*!" We have this chowder recipe as a result, thanks to the John F. Kennedy library archives in Boston.

2	pounds haddock		1	bay leaf, crumbled
2	ounces salt pork, diced		1	teaspoon salt
2	onions, sliced			Freshly ground black pepper
4	large potatoes, diced		1	quart milk
1	cup chopped celery		2	tablespoons butter

1. Put the haddock in a soup pot with 2 cups water and simmer for 15 minutes. Drain, reserving the broth. Check the fish for bones and remove.

2. Sauté the salt pork in the soup pot until crisp. With a slotted spoon, remove the pork and set it aside. Sauté the onions in the pork fat until golden brown. Add the fish, potatoes, celery, bay leaf, salt, and pepper to taste.

3. Pour in the reserved fish broth plus enough boiling water to make 3 cups liquid. Simmer for 30 minutes. Add the milk and butter and simmer for 5 minutes. Serve the chowder sprinkled with the diced pork.

SERVES 6

"Stepping to the kitchen door, I uttered the word 'cod' with great emphasis, and resumed my seat. In a few moments the savoury steam came forth again, but with a different flavor, and in good time a fine cod-chowder was placed before us. . . . Fishiest of all fishy places was the Try Pots, which well-deserved its name; for the pots there were always boiling chowders. Chowder for breakfast, and chowder for dinner, and chowder for supper, till you began to look for fish-bones coming through your clothes."

— HERMAN MELVILLE, WRITING ABOUT A NANTUCKET CHOWDER HOUSE IN *MOBY-DICK* (1851)

THE PILOT CRACKER INCIDENT

When the cook made chowder aboard a square-rigger, the ingredients were standard ship fare: salt pork, fresh or salt cod, and sea biscuits. (Potatoes were added later.) Sea biscuits (also called hardtack) were the forerunner to Crown Pilot crackers (the ones you find in diners in the noisy cellophane bags), a ubiquitous ingredient in a steaming bowl of New England chowder for decades.

Made by Nabisco, the Crown Pilot was the food giant's oldest recipe, acquired when they bought a Newburyport bakery that had been making the recipe since 1792. When demand for the cracker waned in the 1990s, little did Nabisco's bean counters realize the anger they'd unleash among independent-minded, tradition-bound Yankees when they discontinued the cracker (along with 400 other non-performing foods) in 1996. Maybe they also didn't realize what a compelling story it made: a little cracker, beloved by the underdog, abandoned by a multinational food giant. People thought it was a pretty crummy move on Nabisco's part.

Ground zero was Chebeague Island (population: 350) in Maine's Casco Bay, where folks circulated a "Save Our Pilot Cracker" petition. Angry chowder lovers flooded the Nabisco Customer Comment Hotline. Humorist Tim Sample placed a call to CBS's *Sunday Morning*, and soon Maine islanders were venting on national TV, singing "My Bonny Lies over the Ocean" before the cameras with the refrain, "Bring back! Bring back! Bring back my Pilot crackers to me, to me!" As one islander explained, Saltines are fine for sardines, but not for chowder.

Nabisco resumed production in 1997, though their interest was half-baked; after the company was acquired by Kraft, the Pilot cracker was grounded in 2008. Fortunately, Westminster oyster crackers are a reasonable substitute.

Roasted-Corn Chowder with Crab and Bacon

Old Bay seasoning was created in the Chesapeake Bay area in 1939, when creator Gustav Brunn fled Nazi Germany and settled in the region. Old Bay is used quite a bit in Navy ship galleys (probably due to the strong Naval presence in Maryland), and the seasoning is delicious in all kinds of crab dishes and with other seafood as well. It's named after the Old Bay passenger ship that plied the Chesapeake in the early 1900s. In this chowder, you could substitute lobster or fish for the crab.

4	cups fresh corn (about 6 ears)
8	slices bacon
1	celery stalk, diced
1	large yellow onion, chopped
1	pound red potatoes, unpeeled and diced
5	tablespoons butter
5	tablespoons all-purpose flour
6	cups whole milk
3	tablespoons fresh thyme
2	teaspoons salt
	Freshly ground black pepper
½	teaspoon Old Bay seasoning
1	pound fresh crabmeat, flaked or chopped

1. Preheat the broiler.

2. Cut the corn kernels off the cobs and spread the corn in a single layer on a rimmed baking sheet. Broil, shaking the sheet every few minutes, until the kernels are caramel colored, 5 to 7 minutes.

3. Fry the bacon in a pot over medium heat until crispy, about 8 minutes. Remove with a slotted spoon and drain on paper towels; crumble and reserve.

4. Pour off all but 2 tablespoons of bacon fat, and sauté the celery and onion over medium heat until softened, about 4 minutes. Add the potatoes, stir to coat, and then add the butter. After the butter has melted, add the flour, and cook for 2 minutes, stirring frequently. Slowly add the milk, stirring, and reduce the heat to low. When the chowder starts to thicken, add the corn, thyme, salt, a few grinds of pepper, and the Old Bay. Stir to combine and simmer until the potatoes are cooked through, 15 to 20 minutes.

5. Stir in the crabmeat and cook until heated, 2 minutes or so. Ladle into bowls and top with the reserved bacon crumbles.

SERVES 4-6

PREPARING SHRIMP

If your recipe calls for peeling and deveining shrimp before cooking, remove the shells (including the crunchy covering on the shrimp tails) by cracking them with your fingers and pulling them off. Devein the shrimp by running a sharp knife down the back of each shrimp to remove the black streak, and then wash under cold water. Shrimp-cleaning tools that split the shell and remove the vein in one motion are handy if you eat a lot of shrimp.

Shrimp Bisque ~with~ Bourbon

A bisque is a creamy soup made with shellfish; using the shrimp shells in
the cooking process is a classic French technique to extract maximum flavor.
Here, bourbon enhances the shrimp without overpowering it, and rice replaces some
of the cream traditional in bisque. This bisque is a wonderful start to any meal,
as it is not heavy and has a silky mouthfeel.

2	pounds (approximately 33–35) large raw shrimp
2	tablespoons grapeseed oil
	Salt and freshly ground black pepper
¼	cup plus 2 tablespoons bourbon
3	tablespoons unsalted butter
¾	cup chopped celery
½	cup chopped fennel
1	large onion, chopped
1	medium leek (white part only), washed and sliced
2	garlic cloves, chopped
3	tablespoons tomato paste
2	tablespoons all-purpose flour
½	cup white wine
½	cup chicken broth, homemade or low-sodium
5	cups cold water
½	cup basmati rice
2	sprigs fresh thyme
1	bay leaf
½	cup heavy cream
	Salt and white pepper
¼	cup finely chopped fresh chives, for garnish
	Bread or toasted garlic crostini (see page 154), for serving

1. Peel and devein the shrimp (see opposite page), saving all the shells and tails.
   Reserve 16 whole shrimp and coarsely chop the remaining shrimp.

2. Heat 1 tablespoon of the oil in a large soup pot over high heat. Season the
   chopped shrimp with salt and black pepper, add to the pot, and cook until just
   opaque. Remove to a bowl and set aside. Add the remaining tablespoon oil and
   the reserved shrimp shells and tails to the pan and cook, stirring occasionally,
   until the shells begin to brown. Take the pot off the heat and pour the ¼ cup
   bourbon into the pot. Carefully ignite the bourbon with a long kitchen match

continued on next page

or stick flame and let it burn until the flame subsides and the alcohol has burned off. Return the pot to the heat and cook, stirring constantly, until the liquid is reduced by half, about 3 minutes. Transfer the shells and liquid to a separate bowl and set aside.

3. In the same pot, melt 1 tablespoon of the butter over medium heat. Add the celery and fennel and sauté until translucent, about 8 minutes. Add the onion, leek, and garlic, and cook until the onions are soft, about 3 minutes. Stir in the tomato paste and cook until it begins to coat the bottom of the pan and is somewhat caramelized, about 2 minutes. Add an additional 1 tablespoon butter and melt. Add the flour and cook, stirring constantly, to coat all the vegetables, about 1 minute. Add the wine and chicken broth and stir to deglaze the pan, scraping up the browned bits on the bottom with a wooden spoon. Add the water, rice, thyme, bay leaf, and the reserved shrimp shells with all the accumulated liquid. Bring to a boil and then reduce to a low simmer and cover. Cook until the rice is tender, about 30 minutes.

4. When the rice is cooked, remove the thyme sprigs and bay leaf. Push the soup, including the shrimp shells, through a food mill (there are attachments to the Kitchen Aid standing mixer that do this effortlessly) or strain through a mesh strainer, pushing on the shells to extract maximum flavor. Discard the solids and return the liquids to a clean pot set over low heat. Stir in the reserved cooked shrimp, cream, and the 2 tablespoons bourbon, and heat through, taking care not to boil.

5. Melt the remaining 1 tablespoon butter in a medium skillet over high heat and sauté the 16 reserved shrimp seasoned with salt and white pepper to taste, until cooked through, about 3 minutes. (Be careful not to overcook.)

6. Divide the soup among eight bowls. Garnish each serving with 2 whole shrimp and a sprinkle of chives. Accompany with bread or warm toasted crostini. Serve immediately.

SERVES 8

FORGIVING FOOD

Up until the late 1800s, recipes were pretty loose ("one glass of flour" is one of my favorite measurements from an 1825 cookbook). Fannie Farmer of Boston aimed to change the way people cooked at home when she took over as director of the Boston Cooking-School in 1894, and her *Original 1896 Boston Cooking-School Cook Book* (essentially a revised edition of a book written by her predecessor) became a best-seller, emphasizing "scientific knowledge," standardized measurements, and level measures. Recipes became formulas, and her book, which coincided with the rise of the middle class, was enormously influential — women were staying home and wanted to become professional homemakers. Ladies like my great-aunt Elizabeth, who went to Wellesley in the late 1800s and taught school until she married, were given a creative outlet for using their brains at home by elevating domestic chores to a science.

So influential was Farmer that her publisher eventually put her name in the title, though she left the school in 1902 to open Miss Farmer's School of Cookery, a new kind of school aimed at training housewives rather than professional chefs. Ironically, in elevating domestic arts to a science, she inadvertently bolstered the notion that women's work was at home, and intimidated the hell out of us, leading generations of household cooks to abandon a recipe if they didn't have a precise ingredient, or slavishly follow directions to the letter, burying their cooking instinct and trusting someone else's recipe too much ("Why didn't it work? I followed the directions!").

I'm here to tell you that chowders are forgiving — they were made aboard ship and on the beach, after all. If you don't have one fish or shellfish, substitute another. Don't have thyme? Use oregano. Stay loose, and have fun with it. Shake it up.

"I CANNOT NOT SAIL." — E. B. WHITE

Roasted-Corn Soup with Shrimp and Chipotles

This silky, smoky corn-and-shrimp soup has a back-of-the-mouth tingle from the chipotles, and lip heat from the cayenne, but the milk and cream balance the piquancy. With chiles you have the option of cranking up or dialing down the heat by including or discarding the seeds and membrane.

1	head roasted garlic (see box)
1½	cups corn kernels, fresh or frozen
1	tablespoon olive oil
1	medium yellow onion, diced
1	teaspoon salt
¼	teaspoon cayenne pepper
¼	teaspoon coriander seeds

¼	teaspoon cumin
2	cups milk
2	cups heavy cream
1	teaspoon chipotles in adobo, minced
9	large raw shrimp (5 ounces), peeled and chopped
	Chopped scallions, for garnish

1. Preheat the broiler. Squeeze out the cloves of roasted garlic, chop, and set aside.

2. Spread the corn in a single layer on a rimmed baking sheet and broil, shaking the sheet every few minutes, until the kernels are caramel colored, 5 to 7 minutes.

3. Heat the oil in a heavy soup pot over medium heat. Add the onion and garlic and sauté until the onion is transparent, about 5 minutes. Add the salt, cayenne, coriander, cumin, and the reserved corn, and then stir in the milk. Gently bring to a boil, reduce the heat, and stir in the cream. Simmer for 15 minutes.

4. Use a slotted spoon to remove most of the solids (corn and onions) and transfer to a blender. Purée with a little of the cream mixture and then return the corn purée to the pot. Add the chipotles and shrimp, and simmer until the shrimp is cooked, about 10 minutes. Serve hot, garnished with scallions.

SERVES 4

32

ROASTING GARLIC

Preheat the oven to 350°F (180°C). Remove the papery outer husk of a whole head of garlic and slice off the top ¼ inch of the head (making sure to keep the cloves intact). Rub the exterior with olive oil, wrap in foil, and roast for 45 minutes. The garlic is done when it's soft and easy to squeeze. When cooled, squeeze out each clove.

SEA

BAY

BAY VS. SEA SCALLOPS

Bay scallops are smaller than sea scallops (about ½ inch in diameter as opposed to 1½ inches in diameter), and prized for their tenderness and sweetness. Sea scallops are chewier and cheaper — good for a bouillabaise or other stew. Although scallops, like everything else, are available year-round, they come into season during the colder months, from fall through winter.

If you don't bring home your own harvest, ask at the fish counter for "dry" scallops, which are in their natural state. Wet scallops have been soaked in a chemical preservative to extend their shelf life; plumped up with water, they are not as flavorful or delicate, and won't sear as beautifully as dry scallops. Dry scallops have a pure, more concentrated flavor and are mild and delicate.

Creamy Asparagus Soup with Bay Scallops and Frizzled Leeks

Imagine, a creamy soup with no cream. Bay scallops are mild and sweet, and a delightful study in contrasts — firm yet pliant, with a refined texture. While it's not necessary to serve this soup in a shallow bowl, it makes a fetching presentation.

2	tablespoons unsalted butter
1½	pounds asparagus, trimmed and roughly chopped
2	leeks, 1 cut into ½-inch rounds (white and light green parts), 1 reserved for garnish
1	garlic clove, crushed
	Salt and freshly ground black pepper
¼	teaspoon fresh thyme
2	cups chicken stock
2	cups baby spinach
	Vegetable oil
16	bay scallops

1. Melt the butter in a soup pot over medium heat. Add the asparagus, leeks, and garlic, and cook, stirring occasionally, for 3 to 4 minutes.

2. Season with salt, pepper, and thyme, and then add the stock and bring to a boil. Reduce the heat to medium-low, cover, and simmer for 5 minutes. Stir in the spinach and cook for an additional 2 minutes. Transfer the soup to a blender, purée until smooth, pour into a clean saucepan, and keep warm over low heat until ready to serve.

3. To prepare the garnish, slice the white part of the second leek into thin rounds, separating the rings and washing them then patting dry. Pour vegetable oil into a small pan to a depth of ½ inch and heat to 350°F (180°C). Fry the leeks until golden brown and crispy, about 1 minute. Using a slotted spoon, transfer the fried leeks to paper towels to drain. Sprinkle lightly with salt.

4. To cook the scallops, pat them dry and season with salt and pepper. Heat a skillet over medium-high heat, and then add 1 tablespoon of vegetable oil. When the oil is hot, add the scallops, cooking and turning until cooked through and golden, 2 to 3 minutes.

5. Ladle the soup into four shallow bowls and top with 4 scallops each and a garnish of fried leeks.

SERVES 4

Oyster Pan Roast

In New York City, where there are many epicurean attractions, the Grand Central Oyster Bar may be among the finest. Chefs have been cooking oyster pan roasts with style since Grand Central Terminal opened in 1913. On a board above the bar, the names of towns indicate where the oysters hail from — Wellfleet, Wianno, Watch Hill, and other inlets familiar to those who worship the cult of the oyster on the swivel stools at the bar.

An oyster pan roast is an exquisite winter meal — rich, satisfying, a little decadent, with a briny intensity. It's traditional to add celery salt, though I find that the oyster brine makes this dish salty enough. With just six ingredients, each one matters: Worcestershire sauce balances the brine, and the chile sauce, which was exotic in 1913, adds tang, as well as a pinkish tint to the silky stew. It helps to heat your soup bowls before you begin, and use a high-quality clam juice if you don't make your own. Half-and-half is a fine alternative to heavy cream.

16	fresh oysters, in their shells
2	tablespoons unsalted butter
2	tablespoons Heinz chili sauce
1	teaspoon Worcestershire sauce
1	cup heavy cream
2	slices thin white bread
	Sweet paprika or cayenne pepper

1. Shuck the oysters (see page 147), reserving ½ cup of the juice.

2. Melt the butter in a large skillet over medium heat. Add the chili sauce, Worcestershire sauce, cream, and the reserved oyster juice. Simmer over low heat until slightly thickened, 7 to 10 minutes.

3. Add the oysters and simmer uncovered for 5 minutes or until the oysters are cooked through (their edges will begin to curl).

4. While the oysters are cooking, toast the bread, remove the crust, and quarter into toast points. Place the toast points in the bottom of two warmed soup bowls and top with the oysters and the sauce. Dust with paprika and serve immediately.

SERVES 2

CLEANING MUSSELS

Mussels have a bad rap as being finicky, but they are easy to find, clean, and prepare — "sweet onyx jewels," as one fisherman told me. Anchored on rocks, pilings, mud flats, and other mussels, they can be found on islands, inshore bays, and other protected areas. Unlike clams or quahogs, which are best foraged in a bathing suit or wetsuit, mussels can be picked onshore year-round with your shoes on. Mussels, like clams and most bivalves, are sold, caught, and cooked alive.

If you're an armchair forager, go to the fish market and buy mussels the same day you want to cook them. Look for ones that are kept on ice, with glistening shells, which indicates moisture. We've all seen those disgusting dried-up shells, and inside the shriveled dry mussels that look like they've been dead longer than my great-grandparents.

Worse, if they're dried up, they may actually be dead. Avoid a mussel that looks dehydrated, or whose shell is open.

Whether you've bought them or foraged for them, when you get home, put them in a bowl or on a baking sheet covered with a wet towel (the goal is to keep them moist but let them breathe) in the refrigerator. Clean them within an hour of cooking.

If you've bought them at the market, they are clean. If you've foraged for them, you need to clean them — not a big deal. Soak them in fresh water for 20 to 30 minutes (no longer or they'll die), and then drain in a colander. Tear or cut off their "beards" and scrape the shells under cold running water with a stiff brush or knife to remove dirt and barnacles. Discard mussels that are chipped or broken.

Mussels, like clams, will gape open when they're dead. But not all open bivalves are dead. Tap open mussels with another mussel, or try to squeeze them gently shut. If one still stays open, pitch it; it's not faking it.

Bourride *with* Homemade Garlic Aioli

This garlicky fish stew from Provence doesn't rely on any particular type of fish. Use what's fresh and firm — halibut and shrimp, sea bass or red mullet, monkfish and clams — the choice is yours. The garlic mayonnaise (aioli) thickens and enriches this iconic stew. Serve with crusty bread.

2	tablespoons extra-virgin olive oil
2	carrots, peeled and thinly sliced
1	leek, sliced in half then cut into half moons
½	fennel bulb, thinly sliced
4	plum tomatoes, chopped
½	cup white wine
2	cups chicken stock
1	(13-ounce) can Bar Harbor seafood broth or other fish stock
2	tablespoons chopped parsley
3	tablespoons chopped thyme
1	teaspoon salt
2	pinches saffron
6	sea scallops, cut in quarters
½	pound medium shrimp, peeled and deveined (see page 28)
½	pound cod, cut into 2-inch segments
1	pound mussels
¼	cup Aioli (page 40)

1. Heat the oil in a large soup pot over medium heat. Add the carrots, leek, and fennel, and sauté for 5 minutes. Add the tomatoes and cook 2 minutes longer. Stir in the wine and simmer for 5 minutes. Add the chicken stock, seafood broth, parsley, thyme, salt, and saffron, and simmer uncovered for 20 minutes.

2. When ready to serve, with the broth simmering, add the scallops and cook for 2 minutes. Add the shrimp and cod, and cook 2 minutes longer. Add the mussels and cook just until they all open, about 2 minutes. Serve in bowls, topped with a dollop of aioli.

SERVES 4

Aioli

Aioli is a mayonnaise-like Provençal sauce that's delicious with grilled seafood, fried fish, grilled meats, and vegetables. You can change the recipe slightly by adding a tablespoon of chopped fresh herbs at the end — it's awesome either way. Devotees of the traditional mortar-and-pestle method think it creates a beautifully textured sauce, and brings out a richer garlic aroma, but no one will flog you if you're short on time and resort to a food processor or blender (I do!).

½ cup canola oil

½ cup olive oil

1 large garlic clove, finely minced

1 egg

¼ teaspoon dry mustard

¼ teaspoon salt

1 tablespoon lemon juice

Pinch of cayenne

1. Combine the canola and olive oil in a measuring cup.

2. Put the garlic, egg, mustard, salt, lemon juice, and cayenne in a food processor or blender and blend on high, slowly adding the oil mixture until you have the consistency of a beautiful, thick mayonnaise. Refrigerate if not using immediately. Covered, aioli will last a week in the refrigerator.

MAKES 1 CUP

CLEANING CLAMS

If you bought clams at the market, they're clean. But if you bought them from a guy on the dock at Vineyard Haven who had them in a bucket of seawater, ask him if he's cleaned them. And if you've canoed out to a sandy spit in your bathing suit and spent the afternoon at low tide in the muck digging clams, you definitely want to clean them. Now what do you do?

Bring them home in a bucket submerged in several inches of seawater. (If you're in a car driving home and it's scorching hot, put the clams in a cooler and cover with seawater.) Once home, quickly scrub the outside of the shells with a stiff brush under cold running water — fresh water kills clams — to get rid of the mud and grit (or scrub them in a second bucket of seawater). Hard-shelled clams don't have much grit; it's the soft-shelled clams you need to purge, and this is how you do it:

Put them in a container and submerge them again in seawater — either the real thing, or make up a mixture of 3 tablespoons of sea salt (see? sea salt really does come in handy) per liter of water for at least an hour and up to 20 hours (longer than that they'll suffocate and die when the oxygen runs out in the water). If they are particularly gritty, you may want to change the water.

I know this sounds bossy, but use a non-reactive container — galvanized will kill them. Tupperware is good. You want to keep them cool (or close to the temperature of the sea they came from), so depending on the temperature, put them in the refrigerator or a cooler. I put them in a wire basket in the bay and tie them to a mooring near shore. (And you may want to cover the container: clams will spit water at you. They get the last word.)

Give them one final quick rinse before cooking or shucking. Discard any clams that don't "clam up" if tapped.

Kale Soup *with* Linguiça

There's a large Portuguese population in southeastern Massachusetts and
Rhode Island, dating back to the early 1800s, when Cape Verdean and Azorean
immigrants arrived on whaling ships as the industry grew and needed more crew.
By 1870, immigrants from Portugal and the Madeira Islands followed to work in
the mills. The Portuguese culinary influence is felt everywhere — from Portuguese
sweet bread in bakeries to linguiça at the grocery stores to kale soup dished out as
street food at Catholic feasts.

With ingredients similar to other Mediterranean countries (olive oil, onions, garlic,
bay leaf, paprika), the holy trinity of Portuguese cooking is meat, poultry, and seafood.
Linguiça is a smoke-cured Portuguese pork sausage seasoned with garlic, pepper, and
paprika. This soup is wonderful sopped up with hot crusty garlic bread or a Portuguese
sweet roll. You could easily incorporate seafood by tossing in a few handfuls of clams
5 minutes or so before the end of the cooking process.

3	tablespoons grapeseed oil
¼	cup minced garlic
1½	cups chopped yellow onions
8	ounces linguiça, thinly sliced and quartered (or substitute Spanish chorizo sausage, or an Italian sausage with its casing removed, crumbled into small pieces)
8	cups shredded fresh kale, stems discarded
1	pound potatoes, peeled and cubed
8	cups chicken broth
1	teaspoon fresh thyme
1	teaspoon minced fresh rosemary
¼	teaspoon red pepper flakes
	Freshly ground black pepper

1. Heat the oil in a large soup pot over medium heat. Add the garlic, onions,
   and linguiça, and sauté until the onions are soft, about 10 minutes.

2. Add the kale, potatoes, broth, thyme, rosemary, and pepper flakes. Bring
   to a boil, and then reduce to a simmer, and cook, partially covered, until the
   potatoes are tender, 45 to 60 minutes. Season with black pepper to taste
   and serve.

SERVES 6-8

THE COTTAGE

Growing up, we always referred to our summer places as "the cottage." My cousin's cottage was on the WeWeantic, my friend's was on Hamilton Beach, another friend's was on Briarwood Beach, and we all knew whose we were referring to. We never said "our cottage," or "Maureen's cottage," or "Doug's cottage"; it was simply "the cottage."

Even recently, a friend wrote me about someone who visited his mother years ago at "the cottage," and it immediately brought me back to his family's unheated bungalow on a bluff overlooking Parkwood Beach. They haven't owned their cottages in decades, but when my cousin Doug talks about "the cottage" I know he's referring to his family's, and when he asks me about Thanksgiving at "the cottage" this year, I know he's referring to mine.

Seasoned summer places that run deep in a family, sometimes for generations, have a way of doing that. They are so fixed in our kindred memory and spirit that they don't need pronouns to be understood. They become shorthand for a place, a time, a state of mind.

② THINGS THAT SWIM

THIS CHAPTER'S A BIT FISHY. Certainly it was abundant fishing that first brought the Vikings and Basques to Atlantic waters and led early Europeans to form settlements along the coast and islands. Attracted by tales of cod as large as men, explorers came and profited. The rocky shores were suitable for curing cod with salt, which allowed the preserved fish to be transported worldwide. And though this chapter is about all kinds of fish that ply New England waters, cod is arguably the best known, and in some ways, the history of coastal New England can be seen through the lens of the Atlantic cod. The codfather.

A BIT OF HISTORY: when Bartholomew Gosnold arrived in 1602 at what is now Massachusetts and spied a spit of land that Verrazano had named Pallavisino from afar 78 years before (Verrazano didn't even bother to land), Gosnold dropped anchor, went ashore, and found wild strawberries, whortleberries, and plenty of firewood. As noted in his diary, he was greeted by a friendly Indian with copper plates hanging from his ears who "showed us a willingness to help us in our occasions."

Gosnold wrote, "We took great store of codfish, for which we altered the name, and called it Cape Cod." Gosnold pushed on to what is now the island of Cuttyhunk, where he formed a settlement he called Elizabeth's Isle (after the queen), and other merchants followed, drawn by the great opportunity for salt cod, one of the world's greatest sources of protein. (Interestingly, to this day the Portuguese and Spanish have no word for fresh cod — salt cod is *bacalhau* in Portuguese and *bacalao* in Spanish.)

When English Captain John Smith arrived 12 years later in hot pursuit of whales and gold, he found whaling dangerous and a "costly conclusion." But the cod was like gold, and he ended up rich from it. He also mapped the coastline from Penobscot Bay to Cape Cod, noting 25 excellent harbors. It was this map the Pilgrims pored over a few years later, intrigued by the words "Cape Cod," which led them to ask for a land grant to North Virginia, as this region was then called.

"(We were) inclined to go to Plymouth," wrote Massachusetts governor William Bradford, "chiefly for the hope of present profit to be made by the fish that was found in that country."

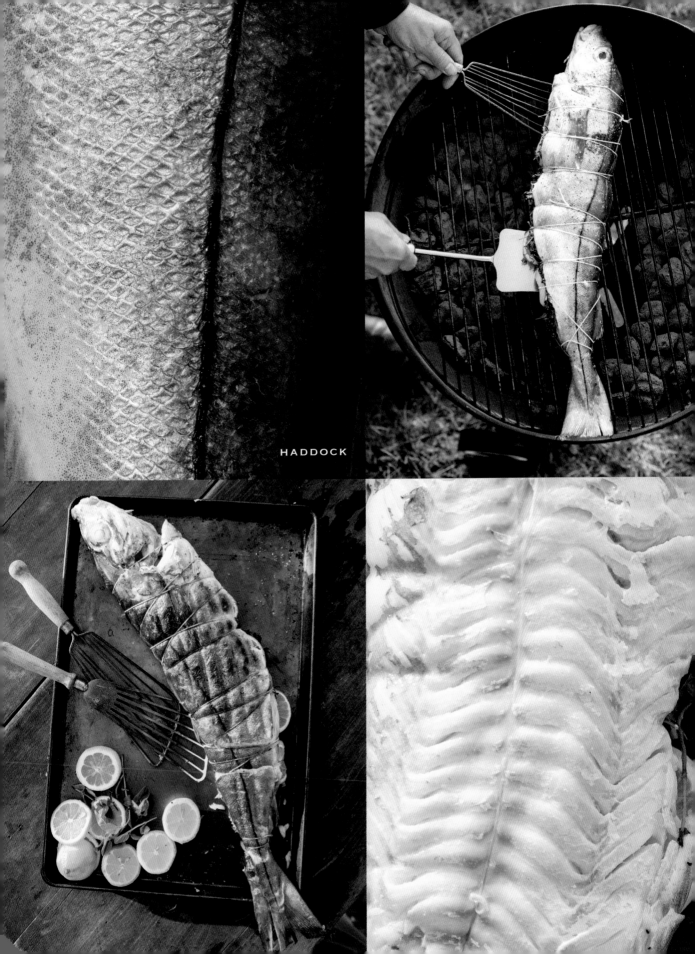

HADDOCK

INDEED YOU CAN'T REALLY
TALK ABOUT NEW ENGLAND SEAFOOD
WITHOUT MENTIONING COD.

There's great irony, or at least chutzpah, in the fact that the Pilgrims knew nothing about fishing, and less about hunting. Arriving on the cusp of winter, they almost starved in one of the world's richest fishing grounds. (Massachusetts even had a winter in-shore fishing season because cod move close to shore for winter spawning.) They didn't know how to fish . . . or hunt . . . or farm, really. Nor did they like unfamiliar food. The Wampanoag taught them to fish (and use cod waste as fertilizer), pry open quahogs, steam clams, and make sense of lobsters. They ate them out of desperation. Mussels they fed to their hogs.

The Pilgrims and other early English colonists did become fishermen, establishing fishing stations in Salem, Gloucester, Dorchester, Marblehead, and Maine's Penobscot Bay. "The abundance of sea fish are almost beyond believing," wrote Salem minister Francis Higgenson in 1629. By 1640, the Massachusetts Bay Colony brought 300,000 cod to the world market.

As early as 1646, a triangulated trade emerged: ships left Massachusetts seaports bound for Europe, where they'd sell high-quality cod and take aboard wine, fruit, etc. Next stop was the West Indies, where they'd sell low-quality cod (to feed slaves) and return home with sugar, spices, coffee, molasses, and cotton. New Englanders imported molasses and made their own rum to sell. By the mid-1760s, there were more than 90 distilleries in Massachusetts and Rhode Island (22 in Newport alone), and so popular was rum in colonial times that the word was a generic term for "alcoholic drink." In the 1700s, ships from New England were loaded with varied cargo, including rum and cod, which they'd take to Africa, then slaves to the West Indies, then sugar and molasses back to New England. The Yankees made money at every stop.

"This is good old Boston, the home of the bean and the cod," goes the ditty, and by the 1700s, a "cod aristocracy" had emerged, with families from Nantucket to Gloucester tracing their prosperity and power in world markets back to shipping. Mansions were decorated with cod. Churches sported gilded cods on weathervanes (smart, this religious and secular motif). To this day, you find codfish on preppy belts, ties, and pocketbooks. In 1776, Adam Smith singled out New England fisheries as a good example of free-trade capitalism. Britain took note; no longer an obscure outpost, the Colonies had become a vibrant commercial entity. The Molasses Act of 1733 followed (a disastrous attempt by the British to assert authority over rum-loving colonists), as well as a succession of acts to hamstring the upstarts. By 1774, with colonists boycotting British goods, King George imposed the 1775 New England Restraining Act, barring colonists from fishing the Grand Banks. Fishing rights were so vital that after the Revolutionary War (and War of 1812), they were a sticking point in negotiations.

Since 1784, a 5-foot wooden Sacred Cod (it's really called that) has hung in the Boston State House, an enduring symbol of cod's importance to the early economy. The speaker of the House faces it in the visitor's gallery, and delegates deliberate under the carved pine sculpture, which, rumor has it, is carved with the phrase *In Cod We Trust*. Traditionally, legislators turned the cod each year, depending on which party was in power.

In a city with 65 colleges, in the annals of college pranks, one of the best involved the cod. In 1933, the Harvard Lampoon staff decided to steal (i.e., cod-nap) the famed sculpture. Pretending to be tourists, three Harvard students walked into the State House on a Wednesday with clippers and a flower box. With no tourists around, they clipped the wires holding the sculpture and disappeared with the fish. They waited, but no one noticed! An anonymous call was placed to reporters, and the fishy business made national news. Journalists had a field day, claiming that several red herring leads were followed, and the Charles River was dredged. Once the Harvard police got involved, it turned up after a few days.

In recent years, cod has become a poster child for commercial overfishing. Severe restrictions on Gulf of Maine Atlantic cod were imposed in late 2014, having a serious impact on the amount of cod available locally, and for good reason. Scientists,

government officials, fishermen, fishery managers, and consumers have the important task of focusing on strategies for long-term sustainability and finding ways to implement those measures. Substitutes for Atlantic cod include Pacific cod, Pacific ling cod, and Alaskan Walleye Pollock.

Meanwhile, the iconic cod is still found on many New England menus, and indeed, you can't really talk about New England seafood without mentioning cod. But cod's just a tiny part of the story; there is a plethora of seafood available in New England, where we pluck our food right from the sea. At the end of the day, whether it's flounder, swordfish, stripers, or blues (bluefish), there's nothing as satisfying as a shore dinner. That's no fluke.

Roasted Cod *WITH* Basil *AND* Tomatoes *ON* Garlic Toasts

When preparing cod, select a fat piece where the center of the fish is 1 inch thick. It should fall apart in cooking and give off milky juices. Beware: if cod doesn't flake, it's not fresh. Nor will it tolerate overcooking — cook it quickly and gently, and prepare it simply. You can substitute haddock, flounder, pollock, halibut, sole, or any thick white fish. This is a terrific dish for a winter dinner party.

1	pound (about 6) plum tomatoes, cored and diced
½	cup finely chopped shallots
4	anchovy fillets, minced
2	garlic cloves, minced
2	tablespoons white wine
½	teaspoon salt
⅛	teaspoon black pepper
4	(4-ounce) cod loins
5	tablespoons grapeseed oil
4	slices Garlic Toast (recipe follows)
½	cup fresh basil chiffonade (see below)
2	tablespoons chopped fresh parsley

1. Preheat the oven to 475°F (240°C).

2. Combine the tomatoes, shallots, anchovies, garlic, wine, salt, and pepper in a medium skillet and simmer over medium heat for 10 minutes.

3. Rub both sides of the fish with the oil and season with additional salt and pepper. Bake in a baking dish or on a baking sheet until just cooked through, 8 to 10 minutes.

4. To serve, place a garlic toast on each dinner plate, top it with the fish, and then a generous serving of the tomato sauce. Garnish with the basil and parsley.

SERVES 4

CHIFFONADE

Chiffonade is a fancy term for thinly sliced strips of flat, leafy herbs or vegetables. It's easy to do: stack the leaves together and roll them, then with a sharp knife, slice perpendicular to the roll to create long, thin strips.

Garlic Toasts

Garlic toast, especially with herb butter, sops up juices beautifully. If you want to cut down on the butter, substitute ¼ cup of extra-virgin olive oil for half the butter. It's a wonderful luxury to have garlic bread in your freezer and be able to use a few slices at a time as needed, so make an extra loaf for when you're short on time. Place the slices on a baking sheet and freeze until the bread is solid, about an hour. Transfer the bread to an airtight container to store until needed.

½	cup (1 stick) butter, softened
⅓	cup freshly grated Parmesan cheese
1	tablespoon minced fresh parsley
2	teaspoons minced garlic
⅛	teaspoon freshly ground black pepper
1	loaf Italian bread, cut into ½-inch slices

1. Preheat the oven to 425°F (220°C).

2. Blend the butter, cheese, parsley, garlic, and pepper in a bowl.

3. Spread the butter on one side of each slice of bread.

4. Bake butter-side-up on a baking sheet until the bread is golden and the butter is melted, about 7 minutes.

MAKES 1 LOAF OF TOASTS

INTERVIEWER: Could you please tell us, in your own words, why did the *Intrepid* lose the race?

COOK: Well, I suppose it was my fault. I started preparing the bouillabaisse as we rounded the last buoy. We were way ahead, and when the soup was ready I called the crew down to the galley. I really don't know what happened after that, except that when we went back up on deck we were sailing in somebody's wake. But on one point everyone's agreed. That was the most fantastic bouillabaisse I've ever made."

— NEIL HOLLANDER AND HARALD MERTES, *THE LAST SAILORS*

FISHY BUSINESS

IT'S 5:30 A.M. and I'm rumbling down Route 195 in a fish truck with Rich Pasquill, who started in the fish business thirty years ago with his dad, and six years later opened a seafood restaurant in southeastern Massachusetts that has the best seafood I've ever tasted. I wanted to learn how he does it.

Both of Rich's grandfathers were fishermen lost at sea — one in a storm off Georges Banks, and the other when a tanker in the wrong lane hit his boat. Their names are on the wall at the Seamen's Bethel in New Bedford, along with other locals lost at sea. "In this same New Bedford there stands a Whaleman's Chapel," wrote Herman Melville about the chapel in *Moby-Dick* more than 150 years ago, "and few are the moody fishermen, shortly bound for the Indian Ocean or Pacific, who fail to make a Sunday visit to the spot."

"No way my father was letting me go to sea," Rich said, backing up to a large metal building on the New Bedford waterfront. We left our lattes in the cab and headed to the morning auction. As we passed the loading dock, Rich explained: for decades his father was in charge of unloading the fish down at the docks, and he learned the trade from him, starting out in high school working on the water boat, a tug that waters the fishing boats.

"So many guys my age ended up on the waterfront because it was flourishing in the '70s," he explained. "Guys were making more than pro hockey players. It was too tempting to not go to college."

The fish auction doesn't begin until 8 A.M., but he's down here early seven days a week, inspecting and sourcing 5,000 pounds of fish weekly. We consider an enormous skate sprawled across shaved ice. Looks good to me. "It's been out in the sun," he says dismissively, flipping it over. "Some of these day-boat guys don't care about fish," he shrugs, pointing out the burn spots.

He's shaking scallop bags and inspecting each box of seafood that was unloaded from 11 P.M. until 4 A.M. this morning in this cavernous warehouse that's spotless and doesn't smell fishy, despite the fact that

WOULDN'T YOU WANT THE FISH THEY CAUGHT YESTERDAY . . .
AS OPPOSED TO WHAT'S BEEN ON ICE FOR A WEEK?

we're up to our gills in it. A guy is hosing down a floor that looks cleaner than my kitchen on a good day.

Deepwater fish is healthier, he confides: the guys who steam out to Georges Bank or Nantucket Sound to trawl for scallops are gone for a week, maybe more. They catch fish every day they're out. Wouldn't you want the fish they caught yesterday, as they headed back to the harbor, as opposed to what's been on ice for a week? He knows these guys, went to high school with them. He asks which load was caught at the end of the trip and buys it.

"Hey Fingers," he nods to a guy sorting fish, then gives me a sheepish look. The guy was missing most of the fingers on his left hand. It's dangerous work all around, from the sorting and processing, to fishing itself, where accidents happen often.

We watch the *Alaska* unload scallops, which have helped make New Bedford famous again. For the last decade, New Bedford has been the most profitable fishing port in the United States due to scalloping. Just a day's steam to Nantucket Shoals, Georges Bank, and the Great South Channel, trawlers bring in 50 million pounds of sea scallops each year — a $411 million per year business.

Rich brings me to the auction room, where an electronic wall display that you'd expect in a commodity broker's office lists fish prices by species and size. Copies of *National Fisherman*

are on the counter along with a pot of old coffee. Two guys look up from their newspapers warily, waiting for the day's auction to begin. "She's with me," Rich answers the silence. On the wall is a sign: "National Marine Fisheries Service — Destroying Fishermen and their Communities since 1976."

After the auction, he heads home. Rich doesn't fish. He doesn't go out on boats. To relax, he sits in a beach chair on his lawn in nearby Mattapoisett, a good distance from the bay. "I have my waterline," he says, smiling.

Baccala (Salt Cod) Salad

I know you're probably thinking, *whaaaat*? Dried salt cod looks like "a last-resort snack for those beyond the Wall on *Game of Thrones*," quipped *Bon Appetit* in naming salt cod the Catch of the Year in 2015, but rehydrated, it's got gumption and a subtle cod flavor. Brought to America by immigrants, this recipe is for an Italian-American baccala salad that my cousin Donna's family prepares every Christmas Eve. Note that you begin preparation several days before serving.

2	pounds salt cod
1	cup extra-virgin olive oil
3	garlic cloves, minced
½	cup lemon juice
1	cup Gaeta or Kalamata olives (oil-cured), pitted and halved
¼	cup capers
2	tablespoons chopped flat-leaf parsley
	Salt and freshly ground black pepper
	Red pepper flakes
	Parsley and lemon wedges, for serving

1. In a nonreactive container, submerge the salt cod in cold water and refrigerate, changing the water every 8 hours for 2 to 3 days to remove most of the salt.

2. Fill a large pot with 8 quarts cold water and add all the cod. Bring to a boil, then reduce to a slow simmer. Cook until the fish breaks apart easily or flakes and is tender, 10 to 15 minutes. It will look opaque, not shiny, and still be sturdy and not soft.

3. Strain and cool until you can handle the fish, and then remove any skin and bones. Break into bite-size pieces with a fork or your fingers. At this point you can store it for a few days, wrapped and refrigerated, but this salad will also hold for several days dressed if you wish to continue.

4. Heat the oil in a medium skillet over medium heat and sauté the garlic for 2 minutes; remove from the heat and allow to cool. Transfer the garlic oil to a bowl and add the fish, lemon juice, olives, capers, and parsley. Add the salt, pepper, and pepper flakes to taste.

5. Transfer the mixture to a serving bowl and adorn with parsley sprigs and lemon wedges. Serve cold or at room temperature.

SERVES 6 AS A SIDE

"He said it must be Friday, the day he could not sell anything except servings of a fish known in Castile as Pollock or in Andalusia as salt cod."

— MIGUEL DE CERVANTES, *DON QUIXOTE*, 1605–1616

NOTE

Olives are easy to pit. Just press down on a few at a time under the flat blade of a chef's knife, and the pits will pop out.

SKATE AND COCKLES

According to Paul Greenburg, author of *American Catch: The Fight for Our Local Seafood*, the world eats 170 billion pounds of fish every year. If everyone ate at least two servings of fish each week, including an oily fish like salmon (recommended for a healthy diet), we'd need 60 billion more pounds to satisfy the demand. Wow. We are accustomed to eating sea bass, cod, tuna, shrimp, and salmon, but we must consider more plentiful alternatives such as skate, mussels, and whelks.

Pretend It's Cod

Let's talk trash. If I called this recipe Crispy Skate, you'd flip the page. Yet skate, which is fished off Georges Bank and the Gulf of Maine, has gone from lobster bait to the rage in Europe, and is just catching on in the United States, mostly in restaurants for now. Resembling a big batwing, skate is a sweet, white, mild-tasting, firm fish that is best cooked simply to enjoy its subtle flavor. In this recipe, you could easily substitute haddock, flounder, pollock, cod, halibut, sole, or most any white fish. This dish goes well with the Arugula and Golden Beet Salad (page 291).

¾	cup panko breadcrumbs
2	tablespoons finely chopped fresh parsley
2	tablespoons finely chopped fresh cilantro
1½	tablespoons extra-virgin olive oil plus more for greasing a baking dish
¼	teaspoon cayenne pepper
½	teaspoon salt
⅛	teaspoon black pepper
4	(4-ounce) skate fillets
	Lemon, for garnish

1. Preheat the oven to 425°F (220°C). Lightly oil a baking dish large enough to hold the fillets in a single layer.

2. Mix the breadcrumbs, parsley, cilantro, oil, cayenne, salt, and pepper together on a plate. Press both sides of the fish into the crumbs and arrange the fillets ½ inch apart in the prepared baking dish, pressing remaining crumbs on top of the fish.

3. Bake for about 17 minutes, until the fish flakes apart when gently prodded with a knife. Serve immediately with lemon wedges.

SERVES 4

Fish Cakes WITH Tomato Sauce

Let's face it: cod can be boring, but this simple recipe, adapted from Yotam Ottolenghi's *Jerusalem*, elevates it remarkably. These cakes can be served with couscous or rice, or alongside a vegetable such as sautéed spinach. If you don't chop the fish up fine, it won't hold together in patties.

TOMATO SAUCE

3	tablespoons olive oil
2	teaspoons cumin
1	teaspoon paprika
1	teaspoon curry powder
1	medium yellow onion, diced small
2	teaspoons minced garlic
1	(14-ounce) can diced tomatoes
½	cup white wine
2	teaspoons sugar
	Salt and freshly ground black pepper

COD CAKES

4	slices white bread
1½	pounds cod, halibut, sole, or other white fish, chopped fine
3	eggs, lightly beaten
½	cup chopped yellow onion
2	garlic cloves, minced
1	tablespoon chopped fresh parsley
1	tablespoon chopped fresh cilantro
1	teaspoon curry powder
1	teaspoon salt
4	tablespoons olive oil
1	tablespoon chopped fresh mint

1. To prepare the sauce: In a large skillet with a lid, heat the oil over medium heat and add the cumin, paprika, and curry powder. Stir briefly, and then add the onions and sauté until soft, about 6 minutes. Add the garlic and sauté for 1 minute, then stir in the tomatoes, wine, sugar, and salt and pepper to taste. Simmer, uncovered, for 10 minutes.

2. To prepare the cakes: Toast the bread and chop with a mezzaluna into a small dice to make breadcrumbs. Combine the bread, fish, eggs, onion, garlic, parsley, cilantro, curry powder, and salt in a large bowl (I find it's easiest to do this with my hands). Shape the cod mixture into cakes ¾ inch thick and 5 inches in diameter. You'll have 8 to 10 cakes. Refrigerate for 10 to 15 minutes to firm up.

3. Heat 2 tablespoons of the oil in a skillet over medium-high heat. Add half the cakes and sear on both sides until browned, about 3 minutes per side. Repeat with the remaining 2 tablespoons oil and cakes. Slide the cakes into the pan with the tomato sauce (they'll be squeezed), add just enough water to cover the cakes partially (no more than 1 cup), cover the pan with a lid, and simmer over the lowest heat for 15 minutes. Turn off the heat and let the cakes settle, uncovered, for 5 to 10 minutes. Serve warm or at room temperature, topped with fresh mint.

SERVES 4

Gone are the days when whaling, shipbuilding, and saltworks supported the coast, but the large docks — long enough for schooners and coasting sloops — still remain in many towns.

Trout Carpaccio

I was served this first course at Las Balsas, a residence turned lakeside inn in the Nahuel Huapi National Park in Argentina. Gabriel, a New York restaurateur and one of my traveling companions, shared the secret of wafer-thin fish carpaccio: freeze the fish before cutting, which makes it easier to slice. By the time you've finished slicing, the fish has thawed.

1	(6-ounce) fillet very fresh sushi-grade trout, salmon, halibut, or other fish

VINAIGRETTE

3	tablespoons extra-virgin olive oil
1½	tablespoons small capers
	Juice of 1 large lemon
1	large garlic clove, minced
½	shallot, minced
½	teaspoon coarse salt
	Freshly ground black pepper
1	cup pickled onions (recipe below)

1. Wrap the fish in plastic wrap and freeze for an hour.

2. To make the vinaigrette: In a measuring cup or small pitcher, whisk together the oil, capers, lemon juice, garlic, shallot, salt, and pepper to taste.

3. Unwrap the fillet and cut wafer-thin slices on a diagonal. Arrange in a single layer around the pickled onions on a serving plate and drizzle with the vinaigrette.

SERVES 2 AS AN APPETIZER

Pickled Onions

So many dishes can be adorned with pickled onions. Use them to brighten a bean soup, embellish a sardine sandwich, or top off a lettuce salad.

1	cup thinly sliced sweet onion
½	cup white distilled vinegar
½	cup water
2	tablespoons sugar
1	teaspoon salt

Put the onion in a medium heatproof bowl. Put the vinegar, water, sugar, and salt in a small saucepan over medium heat. Bring to a boil to dissolve the sugar, and then pour over the onion and cover. Let the onion pickle at room temperature for 1 to 2 hours.

64

Seaweed Sushi Roll

You'd think supermarket sushi would be a dead giveaway: it's not that hard to make. Yet I avoided making sushi for many years out of fear. It took my 12-year-old daughter, Isabel — who got an end-of-semester assignment (four days before Christmas, of course) to make a food from a country she was studying — to get me to take the plunge. She picked sushi, which she loves. So I sighed and said yes and had a blast with her making it. A girlfriend had a tip: watch a short sushi-making video on YouTube before you start, to get the rolling technique down. Then go to a local Asian market, buy dried seaweed (nori) in a packet, sushi rice, and a bamboo mat (maybe $2), and get started. It's easy, you'll impress your guests, and have fun to boot.

If you can't find sushi-quality fish, make it without fish. Sushi is all about the rice, and its complements and presentation. My favorite combination is sweet rice, crunchy vegetables, creamy avocado, and spicy fish.

2	cups sushi rice
6	tablespoons rice wine vinegar
2	tablespoons sugar
5	sheets nori (dried seaweed)
1	cup chopped vegetables (cucumbers, carrots, radishes, roasted peppers, cucumbers, pickled vegetables, cooked spinach, or arugula)
½	cup chopped ripe avocado
4	ounces sushi-grade fish, smoked fish, or cooked crab or lobster or prawns (preferably several fish choices), cut into long skinny strips
	Wasabi
	Soy sauce
	Pickled ginger

1. Rinse the rice in cold water repeatedly, agitating it as you do, until the water isn't cloudy. Drain and cook according to package instructions.

2. In a small saucepan over low heat, combine 4 tablespoons of the vinegar and the sugar. Using a wooden spoon or spatula, turn the rice into a wooden bowl and sprinkle with the vinegar mixture. When your rice has cooled off a bit, you're ready to make sushi.

3. Combine the remaining 2 tablespoons of vinegar with a cup of water in a bowl — you'll use that to dip your fingers when assembling the sushi to prevent the rice from sticking to your hands.

4. Cover your bamboo mat with plastic wrap, and put half a sheet of nori shiny-side down on the wrap. Grab a handful of sushi rice, and spread it evenly over the bottom three-quarters of the nori sheet. Place a thin layer of vegetables, avocado, and fish lengthwise along the center of the rice. (The Japanese suggest an odd number of ingredients — three or five.) Roll the bamboo mat so that the sushi forms a long tube. Don't press too hard. You can roll it into a square, or a roll. Using a good sharp knife, cut into bite-size sushi sections. (If you like, you can turn them on their side and dab each top with wasabi.) Put the sushi on a plate, and serve with soy sauce, pickled ginger, and wasabi.

SERVES 4

A SOLITARY LIFE

The life of a New England fisherman can be a solitary one; many mariners scratch out a living in small boats, often open skiffs of less than 20 feet. From Westport to Eastport, you'll find commercial trawlers often manned by brothers, cousins, maybe a father-in-law visiting from the Azores, or — if you grew up there — some-one you knew from high school.

Fast ∞ Slow

While it sounds greasy, poaching fish in oil is a simple method that yields light, delectable fish. Bear with me through these two simple methods: fast and slow. For the quicker method, choose a thinner fillet, and for the slower preparation, a meatier fillet.

OIL-POACHED FISH: FAST

Sometimes life does not allow the luxury of a lengthy cooking period. This fast method is ready in 15 minutes.

- 3   cups mild olive oil (not extra-virgin)
- 2   pounds flounder, sole, or other thin fillets, brought to room temperature
     Salt and freshly ground black pepper
- 1   bunch flat-leaf parsley, washed and completely dried
- 3   lemons sliced crosswise, paper-thin

1. Pour the oil into a saucepan and heat it to 300°F (150°C) over low heat. (Don't raise the heat; if the olive oil scorches, it will taste bitter.)

2. Preheat the oven to 350°F (180°C) and set a rack in the center. Pat the fish dry and season both sides with salt and pepper.

3. Place half of the parsley sprigs and lemon slices in a cast-iron skillet large enough to hold the fish in a single layer; if necessary, use two pans. Place the fillets on top of the lemons and parsley without overlapping the fish pieces. Cover with the remaining lemon and parsley.

4. When the oil has reached temperature, gently pour it over the fish so as not to disturb. Be sure that all ingredients are submerged in the oil. Place in the preheated oven and poach for 10 minutes.

5. Allow the fish to cool in the pan for 5 minutes before serving.

SERVES 4

OIL-POACHED FISH: SLOW

You can prepare this and serve it right away, or you can make it ahead by cooking, then cooling the fish and refrigerating it in the oil. Just eat it within two days. The oil will keep the fish from drying out and allow it to absorb more flavor as it rests.

4	(6-ounce, 1-inch-thick) portions of halibut, cod, haddock, or scrod, at room temperature
	Salt and freshly ground black pepper
2	cups mild olive oil (not extra-virgin)
¼	cup capers, rinsed and chopped
2	sprigs fresh thyme
3	garlic cloves, peeled and smashed
	Flat-leaf parsley leaves and lemon wedges, for serving

1. Preheat the oven to 250°F (120°C) and set a rack in the center.

2. Pat the fish dry and season both sides with salt and pepper. In an 8-inch glass or ceramic square baking dish, arrange the fish in a single layer, overlapping the thinner areas of the fillets slightly to create a uniform thickness that will ensure even cooking. Pour enough oil into the dish to just cover the fish. You will need slightly more or less oil depending on the size of your dish (use a smaller dish that just fits the fish in a single layer to minimize the amount of oil).

3. Add half of the chopped capers and all of the thyme sprigs and garlic to the oil, tucking them under the surface so they stay moist during the cooking process. Bake, uncovered, until the fish flakes and is cooked through, about 1 hour.

4. Serve by spooning some of the cooking oil over the fish. Sprinkle parsley and the remaining chopped capers over the fish and serve with lemon wedges. The flavorful oil can be used to dress vegetables or rice alongside the fish.

SERVES 4

Bacon-Wrapped Whitefish WITH Charred Fingerling Potatoes AND Frisée Salad

Use any firm, flaky mild fish — cod, haddock, halibut, monkfish, hake, or sea bass. Look for the thickest piece of fresh fish available and make sure you have uniformly thick pieces of bacon. Enough bossiness; this is for your friends who turn their noses up at fish cooked inside — it's surf and turf, the salty bacon hugging the sweet, flaky fish. Yum.

POTATOES

16	fingerling potatoes, halved
2	tablespoons olive oil
2	teaspoons minced fresh thyme
1	teaspoon minced fresh rosemary
1	teaspoon minced garlic
¼	teaspoon coarse sea salt
	Freshly ground black pepper

FISH

2	tablespoons prepared horseradish
2	teaspoons lemon zest
1	teaspoon minced fresh thyme
¼	teaspoon black pepper
1½	pounds thick whitefish loin (of even thickness) cut into 4 pieces
4	slices thick bacon

FRISÉE SALAD

1	grapefruit
4	cups frisée, torn into bite-size pieces
½	small red onion, very thinly sliced
¼	cup chopped and toasted pecans
¾	cup olive oil
1	tablespoon red wine vinegar
1	teaspoon Dijon mustard
½	teaspoon salt
	Freshly ground black pepper

1. Preheat the oven to 475°F (240°C).

2. To make the potatoes: In a large roasting pan (it must be big enough to hold the potatoes and eventually the fish), toss the potatoes with the olive oil, thyme, rosemary, garlic, salt, and pepper. Roast for 15 minutes.

3. While the potatoes are cooking, prepare the fish: Mix the horseradish, lemon zest, thyme, and pepper together. Spread the mixture over the four fish pieces and wrap each piece with a bacon strip.

continued on next page

4. When the potatoes have cooked 15 minutes, make room in the pan for the
fish (tucking the bacon under each loin so the "belt" holds) and roast until the
fish is cooked and the bacon is crisp, about 10 minutes. (If the bacon needs
to crisp up a bit but the fish is done, stick the pan under the broiler for up to
3 minutes.)

5. While the fish is cooking, prepare the salad: Cut the grapefruit in half; peel and
segment one half, and squeeze ⅓ cup juice from the other half. Toss the frisée,
grapefruit segments, red onion, and toasted pecans in a bowl. In a measuring
cup or small jar, whisk the oil, grapefruit juice, vinegar, mustard, salt, and
pepper to taste. Toss the salad with half of the dressing, adding more if desired.

6. Remove the fish and toss the potatoes with the pan juices. Place the potatoes on
a serving platter, top with the fish, and serve with the frisée salad.

SERVES 4

Pesce all'Acqua Pazza

In this Neapolitan recipe ("fish in crazy water"), fish is poached in flavored water — fishermen in Naples would use seawater. It delivers what late summer is all about — fresh fish, tomatoes from the vine, and basil from the garden. The meal is assembled quickly; it's super easy and has loads of flavor. Serve with buttered noodles or couscous.

¼	cup extra-virgin olive oil
3	garlic cloves, chopped
1	small yellow onion, finely chopped
4	medium tomatoes, coarsely chopped, juice reserved
2	tablespoons torn fresh basil leaves plus 8 large leaves for garnish
¼	teaspoon red pepper flakes
½	cup Kalamata olives, pitted and chopped
¾	teaspoon sea salt
	Freshly ground black pepper
2½	cups water
1	cup white wine
4	(5-ounce) fillets white fish (red snapper, cod, haddock, etc.), skin on

1. Heat the oil in a large skillet over medium heat. Add the garlic and onion and sauté until translucent, about 5 minutes. Add the tomatoes, basil, pepper flakes, olives, salt, pepper, water, and wine. Cover and bring to a simmer over medium heat. Simmer for 30 minutes.

2. Uncover the skillet, add the fish (skin side up, to keep in moisture), then cook covered over low heat for 3 minutes. Gently turn the fillets, season with salt and pepper, and simmer uncovered until just cooked through, about 2 minutes.

3. Serve the fish in individual shallow bowls, with tomato broth spooned on top. Garnish with shredded basil leaves.

SERVES 4

Shad Fillets

A sign of spring in New England, just as sure as forsythia blooming in yards and mud on beach roads, is the appearance of shad (also known as herring) at the fish markets and in the weirs as they travel from salt water upriver to spawn. Wampanoags (who called shad "porcupine fish" because of all the bones) taught colonists to cook them slowly over an open fire on cedar planks; the slow cooking dissolves the tiny bones.

A sweet flavorful fish, shad can be cooked simply with a little lemon, salt, and oil. (Indeed, this is a good basic recipe for any broiled fish. You could substitute sea bass, striped bass, red snapper, or other white firm fish.) Put a bunch of forsythia branches in a jar on the kitchen table and serve this dish with some skinny asparagus, lemon rice, and a crisp white chardonnay on an early April night when the days are warm and the nights still cool, where you can look at the water and think "Soon, soon . . . ."

4	shad fillets (approximately ½ pound each)
1	tablespoon extra-virgin olive oil
1	garlic clove, minced
¾	teaspoon fresh thyme
½	teaspoon salt
	Freshly ground pepper
	Butter
	Lemon wedges, for garnish

1. Brush the fish fillets with the oil and sprinkle the garlic, thyme, salt, and pepper to taste over the top.

2. Preheat the broiler and lightly oil a broiler pan. Place the fillets in the pan and dot each with a bit of butter. Broil until golden brown and just cooked through, about 4 minutes. Serve immediately with lemon wedges.

SERVES 4

THE WHOLE TRUTH

You may be cooking fish you caught yourself. Or you may be at the fish market, eyeing one that didn't get away. If you're buying whole fish, here's what to look for:

- Bright shiny eyes (not gray or cloudy)

- Red gills (not the color of faded bricks)

- Flesh that springs back when you poke it (rather than staying indented)

- Glistening skin (not dull or patchy)

- Fish that smells like the sea, not fishy!

Eat fish the day you catch it or buy it. Ideally, store a whole fish in butcher's paper in the coldest part of your refrigerator on a bed of crushed ice (but not directly on the ice).

Wild Striped Bass *with* Pistachio Crust

The reefs off Cuttyhunk can be threatening — the whaling ship *Wanderer* and many other ships have sunk off Sow and Pigs Reef, but the reefs also make Cuttyhunk a prime place for sport-fishing bluefish and striped bass. In 1967, Charles Cinto caught a 73-pound wild striped bass off the island!

½	cup roasted, salted pistachios, shelled
1	tablespoon cornmeal
¾	teaspoon curry powder
¼	teaspoon cayenne pepper
½	teaspoon salt
	Freshly ground black pepper
6	(6–8 ounce) wild striped bass fillets, skin on
2	tablespoons olive oil
1	lemon, cut into 6 wedges

1. Combine the pistachios, cornmeal, curry powder, cayenne, salt, and pepper to taste in the bowl of a food processor and pulse until coarsely ground. Transfer to a flat plate or pan.

2. Coat the flesh side of the bass with the pistachio mixture, pressing it in lightly. In a skillet large enough to hold all the fillets, heat the oil over medium heat. Add the fish and pan-fry, coated-side down, until golden, 1½ to 2 minutes. Flip each fillet carefully and cook the other side until done, about 2 minutes depending on the thickness of the fillet. Transfer to serving plates and squeeze a lemon wedge over each fillet.

SERVES 6

Estrella's Simple Secret

Our El Salvadorean babysitter cooks incredibly tasty dishes — empanadas, rice and beans, fish with lemon — and when I ask her for the recipes, I'm startled by how simple they are, sometimes taking just 10 minutes to prepare. This recipe is good with any white fish that isn't too thin: tilapia, cod, or haddock fillets are perfect.

4   (3- to 4-ounce) fillets (½ inch thick) of any white fish
     Juice of 1 large lemon
     Salt and freshly ground pepper

1. Preheat the oven to 350°F (180°C).

2. Lay the fillets flat in a roasting pan just large enough to contain them without touching. Squeeze lemon over the fish (it's okay if it drips into the pan), season with salt and pepper, and bake for 15 minutes, or until white throughout.

SERVES 4

THE COMEBACK KID

In the rocky outcroppings and reefs along the Connecticut shore, you know it's late May when the anglers are casting in the surf at dawn for striped bass. By June, the stripers will be migrating north up through New England waters, creating an annual biological spectacle as they journey from spawning grounds in the Chesapeake to Maine.

A prize for recreational and commercial fishermen alike, the striper population was in such drastic decline by the 1980s (a combination of habitat destruction, pollution, and overfishing) that in 1984, Congress passed the Atlantic Striped Bass Conservation Act, enforcing limits on the volume that could be caught. Fishermen balked, but it was a groundbreaking moratorium, and it worked; farm-raised striped bass were bred, and by 1995 the population had rebounded and the government was able to ease restrictions for commercial fishermen, and to a limited extent, for recreational fishermen (although it's hard to regulate, even on charter boats). Like most fish stocks, stripers are managed according to how many can be caught without damaging their ability to regenerate. Above 30 percent is considered overfishing.

A sustainable success story, today, despite new challenges — invasive species and warming waters, to name a few — the striped bass is considered one of nature's great comebacks. The Atlantic States Marine Fisheries Commission continues with its fishery management, assessing stock. One thing is for sure: the stripers will return south past Watch Hill in the fall, migrating as ever.

Cuttyhunk Striped Bass ᴡɪᴛʜ Fresh Herbs

With its rich flavor and large firm flakes, striped bass is a popular saltwater fish that is important both for commercial and sport fishing, which has soared in the last decade. Black sea bass and striped bass (aka stripers) are both types of sea bass. Farm-raised bass is a cross between wild striped bass and freshwater white bass. Substitutes include flounder, rainbow smelt, red snapper, rainbow trout, sea trout, or whiting.

4	sea bass fillets, skin on (if possible)
2	tablespoons extra-virgin olive oil
	Sea salt and freshly ground black pepper
2	tablespoons parsley, cilantro, and basil, chopped and combined
	Snips of chives
	Lemon wedges

Prepare a medium-hot fire in a gas or charcoal grill. Lightly brush or spray the fillets with oil, and season with salt and pepper. Place the fish, skin side down, on the grill. (If you have a gas grill, close the cover.) Cook until the flesh is white and flaky, 7 to 10 minutes. Remove the fillets from the grill and garnish with herbs and lemon wedges.

SERVES 4

CUTTYHUNK

MAYBE A MILE AND A half long by a little less than a mile wide, Cuttyhunk is beautifully remote, the outermost island in the Elizabeth chain, a series of 16 islands that resembles a skipped rock on the charts, dividing Vineyard Sound from Buzzards Bay. According to the 2010 census, the population of Cuttyhunk was 52.

The Wampanoags were the earliest settlers, using it as a summer place for hunting and fishing, and then in 1602 Bartholomew Gosnold established there the first English settlement in New England. A group of New York financiers bought most of the island in the 1800s and formed the Cuttyhunk (fishing) Club in 1864; guests included Teddy Roosevelt and Grover Cleveland (who had a place on Buzzards Bay). For years, islanders have made half a living as fishing guides, shellfishers, carpenters, and boatbuilders.

When I was a kid, we'd sail over to Cuttyhunk every summer. There were no hotels on the island and just one ferry that took passengers from New Bedford, so in our little sailboat, we'd have a quiet night with a few other boats in the protected harbor. As the sun was setting over the mizzen, enterprising young men would buzz out to the boats in the harbor in their retrofitted lobsterboats, and with the wave of a hand you could hail a floating raw bar and be served fresh-shucked local oysters on the half shell.

The highest point is Lookout Hill, a glacially formed hill 154 feet above sea level. As kids, we'd scramble up the hill to see the remains of bunkers built by the Coast Guard in 1941 as a lookout for Nazi U-boats, my mother regaling us with stories of what it was like growing up during World War II near the water and having to use blackout curtains even if you lived 20 miles inland. Now it's a picnic area.

A GOOD GRILLING

Grilling fish can be intimidating: it sticks, and once it sticks, it tears apart. And how do you know when it's done? In truth, grilling fish is pretty easy, and a fun way to cook, especially in the warm-weather months, with friends gathered on the deck or in the yard. A few pointers to make it successful:

- Make sure your charcoal or gas grill grate is clean and well-oiled.

- If cooking fillets, pick a thick, uniform cut.

- Brush or spray your fish lightly on both sides with oil and season it before putting it on the grill. (Fish will stick a little on the grill anyway, so don't worry.)

- Fish takes about 10 minutes per inch to cook; to check doneness, separate the middle gently with a fork — it's done when it's opaque inside and flakes easily.

Wood-Roasted Native Striped Bass

Marc DeRego grew up cooking in New England and developed his philosophy of using local and sustainable ingredients in the Pacific Northwest. Back in New England, he worked at the Back Eddy under another notable Westporter, Chris Schlesinger (of Inner Beauty fame, for those of you who know hot sauce). Now head chef and cofounder of Smoke & Pickles, he shared his recipe for cooking a whole striped bass.

1 whole striped bass
Fresh lemon slices and whole herbs such as flat-leaf parsley, chives, or basil, for garnish

DRESSING

Lemon juice

Extra-virgin olive oil

At least 1 handful fresh parsley

At least 1 bunch fresh chives

Sea salt and freshly ground black pepper

1. Using the dull side of a knife, scale one side of the fish by scraping it tail to head, keeping a good grip on the tail. It's best to do this outside. Scaling one side of the fish helps prevent sticking and aids in moisture retention during cooking. Make sure the cavity is clean and the gills are removed.

2. Using a large indirect smoker, bring the smoker up to between 200 and 225°F (100 and 110°C). Place the fish with head or back toward the fire, depending on your smoker's size and setup. Wipe any excess ice or water from the top of the fish, as water can make the skin split if the temperature creeps over 225°F (110°C).

3. Cook the bass until it reaches an internal temperature of 130°F (55°C) at the thickest part, between 1½ and 3 hours depending on the fish's size.

4. While the fish is cooking (or just before serving), prepare the dressing for the fish: Whisk together in a small bowl or pitcher the lemon juice, oil, parsley, chives, salt, and pepper.

5. Using a large pizza peel (baking stone), slide the fish from the smoker and onto a large maple plank for serving. As the fish rests for at least 20 minutes, garnish it with fresh lemon slices and whole herbs.

6. When the fish is ready to serve, pull back the skin. The flesh should be slightly smoky and extremely succulent. Pour some of the dressing on the fish and serve the remaining dressing on the side.

SERVES 2–4

My great-grandfather had a dandy of a boat, a 60-foot Elco motor yacht that he christened the *Louisa II*. The boat was requisitioned by the United States Navy during World War II in an effort to expand the naval fleet quickly for patrol work; my father always told me that the government bought the boats for a dollar, and sold them back to the owners after the war for a dollar. Unfortunately, the *Louisa II* hit a coral reef and sank off Florida before the war was over. All that's left of the boat is the Lenox china, which my great-grandfather had the foresight to remove before giving the boat to the Navy.

Bluefish *with* Lemon-Garlic Mayonnaise

For generations, Cod Kids (a term for those who have grown up on Cape Cod) have caught bluefish and brushed it with mayonnaise to keep it moist while cooking. Mayonnaise is a tasty complement to bluefish, and can be embellished with all manner of herbs and seasonings. Here, it's simply prepared with lemon and garlic, my two favorite "go to" flavors for fish.

1	cup homemade mayonnaise (page 235)
½	teaspoon grated lemon zest
1	tablespoon lemon juice
2	garlic cloves, finely minced
1	tablespoon freshly chopped herbs, such as parsley, cilantro, or chives
8	bluefish fillets, with skin
	Salt and freshly ground black pepper
	Lemon wedges, for garnish

1. Put the mayonnaise in a small bowl and whisk in the lemon zest and juice, the garlic, and the herbs.

2. Preheat the broiler. Lightly oil a baking sheet and lay the bluefish skin side down on the prepared pan. Season the fish with salt and pepper and brush a thin layer of mayonnaise on the top side of each fillet. Use about half the mayonnaise and set aside the rest for serving. Place the pan 4 to 6 inches from the broiler and cook until the fish is cooked through and the mayonnaise is blistering, 8 to 10 minutes, depending on the thickness of the fish. Serve hot with the remaining mayonnaise and the lemon wedges.

SERVES 8

> NOTE
>
> Remove the bloodline from the bluefish before cooking. This is the section of muscle that the fish uses to swim so far.

THE BLUES ARE RUNNING

Along the coast of New England, bluefish arrive in May and run until October, when they head south and offshore as the water temperature drops. Fatty and fine-textured, bluefish should be eaten fresh. (Remove the dark oily strip running through the center of the fish to avoid a "fishy" flavor.)

Most of the larger fish are caught in the fall, when the fish is oilier from its thick coat of winter fat. Blues will put up a fight and are consequently a prized catch for weekend anglers. Some people think bluefish are "too fishy," but blues just need to be eaten fresh; they should be cooked or smoked within 1 to 2 days of being caught. Smaller bluefish are milder and have fewer toxins, which are common in fish with high oil content.

Outdoor Blues

1   (6- to 7-pound) bluefish, cleaned, with head and tail intact
1   lemon, thinly sliced
1   lime, thinly sliced
    Salt and freshly ground black pepper
    Olive oil

1. Open up the fish and line the cavity with lemon and lime slices, and then season generously with salt and pepper inside and out. (You can add chopped fresh herbs to the cavity as well.)

2. Prepare a medium-hot grill or outdoor fire. Coat the outside of the fish with oil and cook over a steady medium heat until the skin is charred and the flesh is firm, 30 to 40 minutes (the general rule is 10 minutes per inch of thickness), flipping halfway through.

SERVES 8–10

BLUE REVOLUTION

So what exactly is aquaculture?

Aquaculture is the breeding, rearing, and harvesting of animals and sea vegetables — fish, fish eggs, crustaceans, mollusks, algae, seaweeds — under controlled conditions. Practiced in water-oriented cultures since ancient times, there's evidence that eels were raised by aboriginal Australians as early as 6000 BC, carp was raised by the Chinese during the Tang dynasty in the 5th century BC, and that ancient Romans cultivated oysters. By the mid-1800s, North Americans began to experiment with aquaculture as well, developing hatcheries as they witnessed the native trout and salmon populations decline.

Today, aquaculture is a thriving business worldwide, with more than 50 percent of the world's seafood consumed by humans coming from aquaculture. According to the Food and Agriculture Organization of the United Nations, aquaculture has grown by about 8 percent per year over the past 30 years, while wild catches have remained constant or declined.

Smoked Chatham Bluefish *with* Capers *and* Sour Cream

It's a big name for an easy appetizer. If I can find smoked fish at my grocery store in the Berkshire hills of western Massachusetts, I bet you can, too, but it's best if you can find bluefish freshly smoked at the fish market near the docks.

24	good-quality crackers or crostini
½	cup smoked, flaked bluefish
4	tablespoons sour cream
2	tablespoons capers
	Juice of ½ lemon

Lay the crackers on a platter and spread a small dollop of smoked bluefish on each. Top the fish with half a spoonful of sour cream, a few capers, and a squirt of lemon juice.

SERVES 6 AS AN APPETIZER

MENEMSHA BLUES

On the western end of Martha's Vineyard, in the village of Menemsha, you'll find Larsen's, a fish market started in 1969 by an offshore fisherman. It's the kind of place where you can get freshly shucked clams and oysters, and smoked bluefish when the blues are running.

If you visit, you may want to stay until sunset — watching the sun set is a Vineyard tradition; some people bring chairs and wine (Menemsha is dry). You can eat on the beach or on the jetty, the kids catching crabs and fishing while you applaud the view, what with bluebloods by land and bluefish by sea.

Smoked Whitefish Pâté WITH Fresh Figs, Gorgonzola Dolce, AND Local Honey

Although I don't think of the highly perishable fig as New England grown, there is a fresh fig cult in Connecticut, where people grow them in their yards, sometimes taking them into the garage for the coldest part of winter. If you can't find them at farmers' markets, they're usually available in grocery stores during the late summer and early fall.

6	ounces softened cream cheese
2	tablespoons lemon juice
2	tablespoons chopped fresh chives
1	tablespoon finely chopped shallots
	Salt and freshly ground black pepper
½	pound smoked whitefish (or smoked salmon), skin removed, fish chopped
10	small fresh figs, halved
2–4	ounces Gorgonzola dolce
1	tablespoon local honey

1. To make the pâté, combine the cream cheese, lemon juice, chives, shallots, and salt and pepper to taste in a bowl and beat until smooth. Add the fish, and stir to combine. Covered and refrigerated, the pâté will last 4 to 5 days. Bring it to room temperature before serving.

2. To assemble the appetizers, place the fig halves skin-side down on a platter. Top each with a tablespoon of pâté, then a teaspoon of Gorgonzola, and a drizzle of honey on top.

MAKES 20 HORS D'OEUVRES

"When your draft exceeds the water's depth you are most assuredly aground." —LINDSAY'S MARITIME LAW

Panko-Crusted Skillet-Cooked Haddock *with* Red Beans *and* Rice

A member of the cod family, haddock is open for fishing year-round, though it is typically caught from May to November in Massachusetts, just like its pollock and cod cousins.

RED BEANS

1	tablespoon olive oil
1	tablespoon minced garlic
½	small yellow onion, diced
2	(15.5-ounce) cans red beans
½	cup white wine or water
1	tablespoon tomato paste
2	teaspoons oregano
	Salt and freshly ground black pepper

HADDOCK

¾	cup panko breadcrumbs
2	tablespoons chopped fresh flat-leaf parsley
2	teaspoons olive oil plus more for skillet
2	teaspoons melted butter
½	teaspoon Martha's Vineyard or other sea salt
¼	teaspoon smoked paprika
¼	teaspoon freshly ground black pepper
4	haddock fillets, approximately ½ pound each
	Lemon wedges
	Hot rice, for serving

1. To make the beans: Heat the olive oil over medium-low heat in a large skillet. Add the garlic and sauté until fragrant, about 1 minute. Add the onion and sauté over medium heat until soft, about 10 minutes. Add the beans, wine, tomato paste, and oregano, stirring to combine. Lower the heat and simmer for 20 to 30 minutes while you start the fish.

2. To prepare the fish: Preheat the oven to 425°F (220°C). In a dry cast-iron skillet, spread the breadcrumbs in a single layer and bake until golden brown, about 5 minutes, stirring once or twice. Place the crumbs in a wide, shallow bowl and toss with the parsley, oil, butter, salt, paprika, and pepper. Brush the skillet with oil. Press both sides of each fillet into the crumbs, and then lay the fish flat in the skillet. Press any remaining crumbs on top of the haddock. Bake 10 to 12 minutes, or until the fish flakes when lifted with a fork. Serve on a plate with lemon wedges.

3. Stir the beans one last time, season with salt and pepper to taste, and serve with rice and the fish.

SERVES 4

No-Flip Flounder with White Bean Ragout and Spinach Salad

Halibut is the largest of the "flat fish," or flounder, family, reaching upward of 400 pounds. Everyone's probably had the experience of flipping fillets that broke apart into a mess. If you don't want to flip out, pan-fry on one side to crisp the skin and then finish the cooking in the oven. Fillets under an inch thick get the best results, but this technique also works with thicker cuts of halibut or cod. The ragout and salad fill out an otherwise singular presentation.

BEAN RAGOUT

2	cups dried Great Northern beans
2	quarts cold water
1	bay leaf
1	sprig thyme
3	garlic cloves
	Salt and freshly ground black pepper

FISH

4	(6-ounce) halibut fillets, with skin
	Salt and freshly ground black pepper
4	tablespoons neutral oil (sunflower or vegetable) for pan-roasting
	Sea salt

SPINACH SALAD

¼	cup pine nuts
8	slices bacon, diced
½	cup olive oil
2	medium shallots, sliced in rings
¼	cup balsamic vinegar
8	ounces baby spinach, washed and dried
	Salt and freshly ground black pepper

1. Rinse the beans in a colander and place in a large bowl. Cover with 2 inches of cold water and soak 12 hours or overnight in the refrigerator. Drain and rinse well.

2. To prepare the bean ragout: Place the soaked beans in a large pot, cover with 2 quarts cold water, and add the bay leaf, thyme, and garlic. Bring to a slow boil over medium heat; boil for 5 minutes, and then reduce the heat to low and simmer until the beans are tender, usually 1 to 1½ hours.

3. Transfer half of the beans and the remaining cooking liquid to a blender. Remove the stopper from the lid to let steam escape, and place a dishtowel over the top to prevent a hot mess. Return the purée to the pot and mix with the whole beans. Season with salt and pepper to taste and keep warm.

4. To prepare the spinach salad: Toast the pine nuts in a small, dry skillet over medium heat until golden, about 2 minutes. Remove from the pan and set aside. In the same skillet, cook the bacon in olive oil over medium heat until thoroughly cooked, about 4 minutes. Add the shallots and sauté until crisp, about 1 minute. Watch carefully, and do not allow the shallots to brown. Remove the bacon and shallots with a slotted spoon and reserve separately from the pine nuts. When the oil has cooled for 2 minutes, whisk in the vinegar. Season with salt and pepper to taste and keep warm. Wash and dry the spinach, and set aside to keep cool.

5. Preheat the oven to 425°F (220°C).

6. Remove any small bones from the halibut fillets and pat the fish dry. Season with salt and pepper on both sides. In a large, oven-safe skillet, heat the oil over medium-high heat until hot and almost smoking. Place the fish, skin-side down, in the pan. Direct the fish away from you in the oil to prevent oil splashes. Do not let the fillets touch or overcrowd the pan. Sear for 3 minutes without disturbing.

7. Place the skillet in the oven on the top rack for about 7 minutes to finish cooking. If your fillets are thick, they might need an extra minute in the oven. Remove the skillet from the oven and allow it to sit for a few minutes. The fish will continue to gently cook. It is done when it is opaque and has an internal temperature of 145°F (63°C).

8. Toss the spinach in a bowl with the pine nuts and 2 to 3 tablespoons of the dressing. Spoon the beans onto the middle of each plate and, using a thin fish spatula, gently place the halibut, skin-side down, on the plated beans. Top each serving with spinach and drizzle more warm dressing over all. Garnish with the reserved shallot rings and bacon pieces. Finish with a light sprinkling of sea salt.

SERVES 4

Baked Haddock Fillets *with* Horseradish-Chive Potato Mash

This is a simple preparation, good year-round, but especially in deep winter when it's too cold to go out on the deck and fire up the grill. You could substitute any firm white fish such as cod, halibut, sole, or grouper.

4	haddock fillets, approximately ½ pound each
	Juice of ½ lime
1	tablespoon mayonnaise
⅛	teaspoon fresh minced garlic
⅛	teaspoon salt
	Freshly ground black pepper
½	cup fresh breadcrumbs
1½	tablespoons butter, melted

HORSERADISH-CHIVE POTATO MASH

1½	pounds Yukon gold potatoes, peeled and quartered
1	tablespoon extra-virgin olive oil
⅓	cup sour cream
1	tablespoon prepared horseradish
2	teaspoons chopped chives
¼	teaspoon salt

1. Preheat the oven to 425°F (220°C).

2. Place the fish fillets flat in a buttered baking dish. Whisk the lime juice, mayonnaise, garlic, salt, and pepper in a small bowl and brush the mixture over the fish. Sprinkle with breadcrumbs, then drizzle with butter. Bake until the fish flakes when tested with a fork, about 20 minutes.

3. To make the potatoes: Put the potatoes in a large pot of salted water and bring to a boil. Simmer until tender, about 20 minutes. Drain, return to the pot, and add the oil, sour cream, horseradish, chives, and salt. Mash until smooth.

SERVES 4

94

"I had awakened at five and decided to fish for a few hours. I rowed the dinghy out to the boat on that lovely foggy morning and then headed around my side of Martha's Vineyard into the heavy waters of West Chop. Up towards Lake Tashmoo I found the quiet rip where the flounders had been running, put out two lines, and made myself some coffee. I am always child-happy when I am alone in a boat."

— LILLIAN HELLMAN

Flounder Fillets

This is a good winter dish to bake in the oven — it cooks quickly, doesn't leave a smelly residue, and is easy enough to prepare for company. Pop it into the oven during cocktails. Any white fish will do. In terms of vegetables, use what's in the refrigerator: mushrooms, yellow peppers, etc. It's a flexible, healthy, tasty dish.

2	tablespoons olive oil
1	cup sliced yellow or white onion
1	cup julienned red bell pepper
½	cup chopped celery
2	garlic cloves, minced
2	teaspoons fresh thyme or parsley
	Salt and freshly ground black pepper
1	cup chopped tomatoes
4	flounder fillets, approximately ½ pound each

1. Preheat the oven to 500°F (250°C).

2. Heat the oil in a large skillet over medium heat. Add the onion, bell pepper, celery, and garlic, and sauté until soft. Add the thyme and salt and pepper to taste. Add the tomatoes and cook, stirring, for 5 to 10 minutes, or until the juices evaporate a little. You don't want the mixture to be runny.

3. Spread four sheets of foil, each roughly 12 by 12 inches, on the counter and spray lightly with cooking spray. Lay each fillet flat on a foil sheet, placing it halfway down the sheet. Top each fillet with the onion mixture and fold the other half of the foil over the fish. Crimp all three sides, taking care to get the air out of the pouch.

4. Lay the foil packets flat on a baking sheet and bake for 12 to 14 minutes, depending on the thickness of the fillet. Put a packet on each plate and let diners open the foil at the table.

SERVES 4

Chipotle Fish Tacos *with* Cilantro

Setting sail as the wind breezed up after lunch, we'd head to the Vineyard, Naushon, Cuttyhunk, or one of the other Elizabeth Islands. After a day of brisk sailing, we'd toss anchor in a protected harbor. Someone would drop a fishing line off the stern of the boat and, if lucky, catch a fish. Hauling up a flounder, Dad would say "Dinner!" He'd cook the fish simply: dredged in flour and sautéed in butter, with garlic if we had any, and lemon. One small flounder will serve two prepared as tacos.

4	(8-inch) flour tortillas
¼	cup mayonnaise
1½	teaspoons chipotle in adobo sauce
1½	teaspoons lemon juice
2	teaspoons fresh chopped cilantro
¼	head savoy cabbage, thinly sliced
½	small jalapeño, seeded and thinly sliced
½	cup panko breadcrumbs
	Salt and freshly ground black pepper
½	cup vegetable oil
8	ounces flounder fillet, cut into long strips
1	small tomato, chopped
	Lime wedges (optional)
	Sour cream (optional)
	Salsa (optional)

1. Preheat the oven to 350°F (180°C).

2. Wrap the tortillas in aluminum foil and warm in the hot oven for 8 minutes.

3. In a medium bowl, whisk together the mayonnaise, chipotle sauce, lemon juice, and cilantro. Add the cabbage and jalapeño, and stir to blend.

4. Combine the breadcrumbs, salt, and pepper on a plate. Pat the flounder dry with paper towels and dredge through the crumbs, pushing them into the fish. Heat the oil in a skillet over medium heat. Add the fish and fry over medium heat, flipping once, until golden brown and crunchy on both sides. Set aside on paper towels.

5. Working quickly, put a portion of the fish, along with the cabbage mixture, in a warm tortilla and top with chopped tomato. Garnish with a squeeze of lime, sour cream, and salsa, if desired.

SERVES 2

BLUEFISH

FLOUNDER

HADDOCK

SALMON

FISH FOOD

Fish is brain food. It's easy to digest, and it's a low-fat source of high-quality protein. Most seafood contains omega-3 fatty acids (obtained from algae and phytoplankton), which fight chronic disease and are important to a healthy diet. The American Heart Association recommends that adults eat at least two 3- to 6-ounce servings of seafood per week, which not only helps your heart, but your brain, circulation, teeth, skin, eyes, and immune system. The fish in our cold waters — salmon, tuna, halibut — are particularly good. Be smart, eat more fish!

Gefilte Fish

Sometimes called a pescatarian meatloaf, gefilte fish is an Ashkenazi Jewish appetizer made from a mixture of ground, boned fish such as whitefish, pike, or carp. This popular culinary staple became a punchline for Catskills comedians: ("What kind of cigarettes do Jewish mothers smoke?" "Gefiltered.") My cousin Donna's grandmother remembers fish swimming in her bathtub before the holidays at her apartment in the Bronx. To prepare the holiday meal, her mother would grind the flesh of a freshwater fish with other ingredients to form the patties.

3	small onions, 2 peeled and quartered, 1 sliced in rings
3	carrots, 2 peeled and quartered, 1 peeled and sliced on an angle
2	celery stalks, quartered
2	(5-ounce) cans white albacore tuna, drained
1	pound fresh, uncooked whitefish, such as carp or pike fillets, skin and bones removed
4	eggs
2	tablespoons sugar (optional)
	Salt and freshly ground black pepper
½–¾	cup matzo meal, or more if necessary
	About 4 quarts cold water, or enough to just cover
	Prepared horseradish, for serving, preferably the purple kind, with beet juice

1. Place the quartered onions, quartered carrots, celery, tuna, and whitefish in a food processor and pulse a few times until you have a uniform mixture with visible bits of each item. Do not overprocess. Add the eggs and pulse again. If you prefer it sweet, as in southern Poland, add the sugar. Add salt and pepper and enough matzo meal to make a somewhat stiff consistency. It will be sticky but hold its shape. Refrigerate for 1 hour.

2. Fill a large pot with a tight-fitting lid with water and bring to a boil. Add 2 teaspoons of salt, the onion slices, and the sliced carrots, bring to a boil, then reduce the heat and leave the carrots and onions to simmer. You will use the carrot slices as a garnish later.

3. Remove the fish mixture from the refrigerator and make a mini test patty. Put the patty in the simmering water; if it holds its shape, then the mixture is the right consistency. If it breaks apart, add more matzo meal to stiffen the fish mixture.

continued on next page

4. When ready to cook, form the patties into oval shapes, about 3 inches long. Gently place them in the simmering water, touching but not on top of one another. Be sure that all the patties are submerged in the water; add liquid if necessary to just cover. Simmer, covered, for 45 minutes. A rolling boil may agitate the patties, causing them to break apart. Gently shake the pot occasionally to prevent sticking.

5. Remove the pot from the heat and let the fish patties rest in the water for 15 minutes. Use a slotted spoon to carefully remove the patties and drain.

6. Arrange the fish patties on a platter and adorn each with a sliced carrot for decoration and tradition.

7. Refrigerate until cooled. The patties can be held in the refrigerator for a few days. Serve with prepared horseradish.

SERVES 8 AS AN APPETIZER

100

"My grandfather once told her if you couldn't read with cold feet, there wouldn't be a literate soul in the state of Maine."

— MARILYNNE ROBINSON

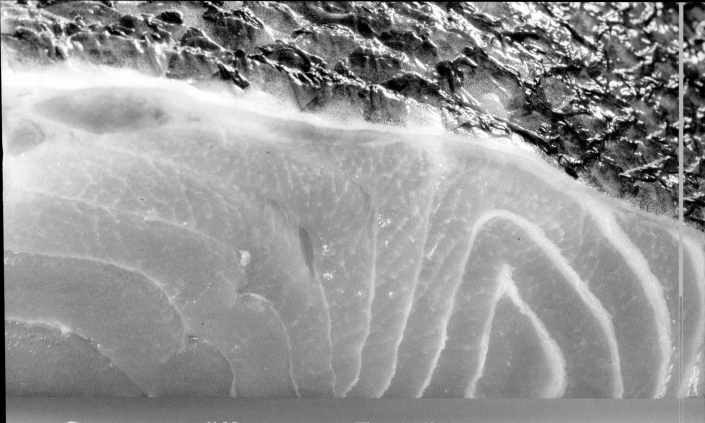

SALMON, "KING OF FISH"

Atlantic salmon are intrepid voyagers, making an epic journey to migrate from freshwater streams and rivers out to the North Atlantic to feed and back up New England rivers in the spring to spawn where they were hatched. An Atlantic salmon might be born in a Maine river in May and found off Newfoundland by late July.

Easy to catch and delicious to eat, they were caught by Native Americans and Europeans, but continued to flourish through the 1700s. In colonial times, Atlantic salmon swam in every river north of the Hudson, and the "king of fish" became a symbol of abundant wilderness, swimming in clear streams. As early as the 1800s, however, they became extinct in several big rivers (the Connecticut and the Merrimack, among others) due to pollution from textile mills, overfishing, timber cutting, and weirs and dams that blocked their natural pathways.

Today, although the industry is improving, environmental impacts are still a problem for this popular fish. Overfishing, acid rain, environmental changes, dams, warming sea-surface temperatures in the Gulf of Maine, and other disturbances have reduced wild salmon stock to dangerously low levels. They remain extinct in 84 percent of New England rivers, and commercial fishing for the species is prohibited, although effort is being made to restore them to their natural habitat.

High in heart-loving omega-3 fatty acids, Atlantic salmon have been raised in hatcheries since 1864, but have returned to Maine more recently through aquaculture, where they are spawned in freshwater hatcheries and grown out in floating net pens alongshore. Wild salmon is another choice — richer and more deeply flavored than farmed. In the absence of wild Atlantic salmon, choose wild Alaskan salmon in season (May through September, when they spawn), which are fit, fat, and flavorful before charging upstream to reproduce. Out of season, the best choice is frozen wild salmon.

Grilled Salmon with Tomato-Basil Relish

Salmon is a full-flavored oily fish loaded with protein and omega-3 fatty acids. You could substitute mackerel, bluefish, herring, or striped bass in this recipe. Or choose wild Alaskan salmon.

SALMON

2	pounds Atlantic salmon, preferably with skin still on one side
2	teaspoons sesame oil
¼	teaspoon dried rosemary
	Salt and freshly ground black pepper
	Lemon wedge

RELISH

2	ripe tomatoes, diced
1	tablespoon extra-virgin olive oil
1	tablespoon balsamic vinegar
1	tablespoon minced basil
1	teaspoon minced garlic
½	teaspoon cumin
	Salt and freshly ground black pepper

1. Prepare a medium-high fire in a gas or charcoal grill. Brush both sides of the fish with the sesame oil, and then sprinkle the flesh side with the rosemary, and salt and pepper to taste. Squeeze the lemon over the fish and grill until it flakes with a fork, 4 to 6 minutes per side.

2. While the fish is cooking, assemble the relish: Combine the tomatoes, olive oil, vinegar, basil, garlic, and cumin in a small bowl. Stir and season with salt and pepper to taste.

3. Serve the fish hot off the grill with the tomato relish on the side.

SERVES 4

BIG FISH

You've caught a fish and boned it, and now you have an enormous fillet. (Ok, maybe you've gotten a nice big one at the market.) No matter — grilling a big 3½-pound fillet is ridiculously easy. I did it in the mountains on New Year's Eve, and my friends looked aghast, coming at me with a cutting board and knife to chop it into neat fillets before I headed to the grill on the porch. Don't bother (and don't remove the skin either). Whether it's halibut, salmon, or steelhead trout, simply brush both sides of the fish lightly with olive oil (or grapeseed oil, or whatever is handy), spray your grill grate with a nonstick oil, prepare the grill, and cook the fish 5 minutes per side, grill cover down, or until the fish is opaque and flakes easily. It's that easy.

Salmon Tacos

My friend Monica Helm has a terrific palate. Combine that with a zest and critical inquiry that she applies to most subjects that cross her path, and you've got a terrific recipe developer. We were shooting the breeze one day over lunch, and she told me about this awesome taco recipe, which she perfected with her brother.

2	cups shredded red cabbage
¼	cup mirin
½	cup mayonnaise
1	teaspoon Tom Yum paste
1	teaspoon lemon juice
8	(10-inch) flour tortillas
1½	pounds salmon, cut into ½-inch strips
1	(13.5-ounce) can coconut milk
2	cups panko breadcrumbs
¼	cup olive oil
2	ripe Hass avocados, thinly sliced
1	lemon
	Salt

1. Preheat the oven to 350°F (180°C).

2. Toss the cabbage with the mirin in a small bowl.

3. Whisk together the mayonnaise, Tom Yum paste, and lemon juice in another bowl.

4. Wrap the tortillas in aluminum foil and warm in the oven for 10 minutes.

5. Dip the salmon in coconut milk, then roll in the breadcrumbs. Put the oil in a skillet and heat over medium heat, until rippling, then sear the salmon on each side until golden brown and crisp, 2 to 3 minutes per side. Transfer to paper towels.

6. To assemble each taco: Lay the tortilla flat and spoon a thin layer of the spicy mayonnaise on one side. Add a portion of cabbage, salmon, and 3 thin slices of avocado. Squirt with lemon, sprinkle with salt, then roll up and eat.

SERVES 4

Atlantic Salmon *with* Grilled Corn *and* Upland Cress Salad

Let's face it, fish can be intimidating. Some of it looks darn weird. We worry about mislabeling. It's easy to overcook. It's as bad as company after three days. Fear not with salmon — it's the poster child of the sea. Salmon has oily flesh that stays moist when cooked. Thick and suitable for filleting, it converts even the most suspicious meat eaters. It's a pretty pink color. It's incredibly good for you. And the skin is edible — crispy salmon skin is like a potato chip from the sea. Just remember to scale it.

3	small ears corn, shucked
1	tablespoon basil chiffonade (see page 53)
½	small red onion, thinly sliced
½	teaspoon salt
2	tablespoons rice wine vinegar
5	tablespoons extra-virgin olive oil
1	tablespoon sesame oil
2	teaspoons pickled ginger, finely chopped
2	bunches upland cress (about 4 ounces total) or arugula
4	(6-ounce) salmon steaks, skin on
	Grapeseed oil
	Salt and freshly ground black pepper

1. Prepare a medium-hot fire in a gas or charcoal grill. Roast the corn until charred, turning frequently, 7 to 8 minutes. Let cool and then, using a serrated knife, cut the kernels off the cobs.

2. Toss the corn in a bowl with the basil, onion, salt, vinegar, olive oil, sesame oil, and the pickled ginger. Add the cress and toss gently to coat.

3. Reheat the grill. Brush the salmon with grapeseed oil, season with salt and pepper, and grill over medium-high heat, covered, for 5 to 7 minutes per side. Divide the salad among four plates and top with the salmon.

SERVES 4

Cedar-Planked Salmon

A plank of salmon makes a beautiful statement and is easy to cook. The soaked plank releases moisture, causing the fish to steam gently. Moreover, the fragrant cedar smoke helps cure the outside (further locking in the moisture) and the salmon will pick up the woodsy flavors from the cedar. As a further enhancement, try soaking the plank in wine or cider instead of water. Be sure to use an untreated cedar plank, available at many hardware stores. Planked salmon is tasty served with Old-Fashioned Cucumber Salad (page 260).

4	cedar planks
2	tablespoons olive oil
	Bunch of fresh dill
2	lemons, thinly sliced
4	(6-ounce) salmon steaks
	Salt and freshly ground black pepper

1. Soak the cedar planks in water for 2 hours. Remove and wipe down.

2. Prepare a medium-hot fire in a gas or charcoal grill. Brush one side of each plank with olive oil. Place a few dill sprigs in the center of each plank (on the side with the olive oil), then 3 slices of lemon, then the salmon. Season with salt and pepper to taste. Place 2 more sprigs of dill on top, and then a few more slices of lemon.

3. Lay the planks on the rack of the grill, turn the grill down to medium-low, and cook for 20 minutes, covered. While cooking, check the cedar periodically; it will burn around the perimeter, but should not burn where the fish is. (It's good to keep a squirt gun handy in case the salmon catches fire — or your cat gets ambitious.)

4. For a dramatic presentation, bring the salmon on the planks to the table. Serve with a bowl of cucumber salad.

SERVES 4

CEDAR-PLANKED SALMON AND SPICY CUCUMBER SALAD

Crispy Roasted Fish WITH Gingered Fig Chutney

You've reeled in a good-looking snapper, flounder, or striped bass. Now what do you do with it? Roasting it whole, with the bones in, will make the meat even more succulent and infuse the fish with flavor. (Think of the difference between grilling boneless chicken breasts and those with the bone intact.) Serve with Gingered Fig Chutney (recipe on facing page).

4	(1- to 1½-pound) whole fish, cleaned, heads and tails intact
2	tablespoons extra-virgin olive oil
2	garlic cloves, minced
	Salt and freshly ground black pepper
2	ripe tomatoes, thinly sliced
2	lemons, thinly sliced
2	cups fresh herbs, such as oregano, thyme, basil, or mint
4	bay leaves

1. Preheat the oven to 425°F (220°C). Cut three crosswise slits (½ inch deep) through the skin on one side of each of the fish. Pat the fish dry. Using aluminum foil, line a shallow baking pan large enough to hold the fish, and arrange the fish next to each other in it.

2. Combine the olive oil and garlic and brush on the fish, inside and out. Season inside and out with salt and pepper. Place the tomatoes, half of the lemon slices, most of the fresh herbs, and a bay leaf in each of the fish cavities. Place the remaining lemon slices and herbs on top of the fish, and roast, uncovered, until the skin is crisp and the fish flakes easily with a fork, about 30 minutes.

SERVES 4

Gingered Fig Chutney

I know a woman who grows subtropical fig trees in Connecticut and enjoys their sweet fruit each fall. She doesn't even bring the plants indoors in the winter — she just bundles them up in a winter wrap. This chutney is a good accompaniment to roasted fish (and pork or lamb), it's delicious spread on a turkey and cheese sandwich, and it's so good that I've been known to stand in front of an open fridge and spoon a little right out of the jar.

½	red onion, coarsely chopped
4	tablespoons coarsely chopped fresh ginger
2	jalapeño peppers, seeded and coarsely chopped
	Juice of ½ lemon
1	teaspoon grated lemon zest
¼	cup raisins
¼	cup chopped apricots or other diced dried fruit
⅔	cup apple cider vinegar
⅔	cup firmly packed brown sugar
2	teaspoons ground cardamom
½	teaspoon salt
¼	teaspoon cayenne or other red pepper powder
4	fresh figs, peeled and cut into ¼-inch dice

1. Combine the onion, ginger, and jalapeños in the bowl of a food processor and blend until the pieces in the mixture are the size of rice. Spoon the mixture into a skillet and add the lemon juice, lemon zest, raisins, apricots, vinegar, brown sugar, cardamom, salt, and cayenne, and cook over low heat for 1 to 2 minutes.

2. Add the figs, bring the mixture to a slow boil, and stir until the juices thicken. Remove from the heat and allow the chutney to cool completely. Refrigerate before serving.

MAKES 1 CUP

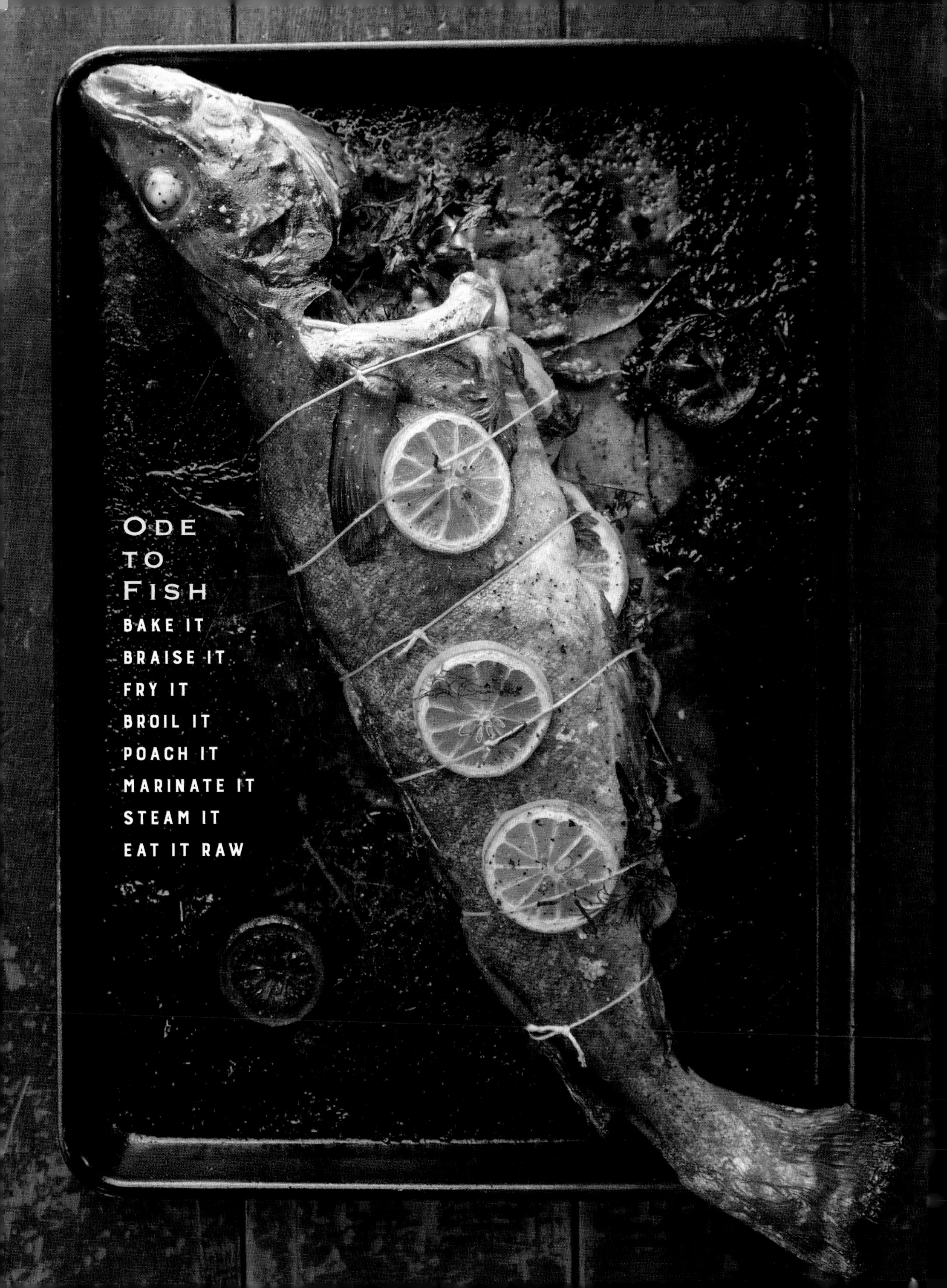

ODE to FISH

BAKE IT

BRAISE IT

FRY IT

BROIL IT

POACH IT

MARINATE IT

STEAM IT

EAT IT RAW

Grilled Shark *WITH* Avocado–Sweet Corn Relish

Upon graduating from college, unlike most of my career-driven classmates, I moved to Mount Desert Island and waitressed for a few months. As the cold weather descended and the tourists departed, the boats started heading south for the winter. When asked if I'd like to crew on a boat delivery south, I didn't hesitate. The second day out, the captain reeled in a Cape shark, which we grilled off the stern that night. The fish in English "fish and chips," Cape sharks are delicious, with lean white meat, a firm texture, and a sweet mild flavor. Halibut is another good choice for this recipe, or you can substitute hake, flounder, salmon, swordfish, or pollock.

RELISH

2	ripe Hass avocados, cut into bite-sized pieces
1	cup cooked corn
¾	cup cherry tomatoes, quartered
½	small red onion, diced
½	jalapeño, seeded and diced
2	tablespoons prepared salsa
2	tablespoons chopped cilantro
2	tablespoons extra-virgin olive oil
1	tablespoon lemon juice
1	teaspoon salt
1	teaspoon minced garlic
½	teaspoon cumin
	Freshly ground black pepper

4	(4-ounce) shark or halibut fillets
1	tablespoon grapeseed oil
	Salt and freshly ground black pepper

1. To make the relish: Combine the avocado, corn, tomatoes, onion, jalapeño, salsa, cilantro, olive oil, lemon juice, salt, garlic, cumin, and pepper in a bowl.

2. Prepare a hot fire in a gas or charcoal grill and oil the grate. Rub the shark with grapeseed oil and season with salt and pepper on both sides. Cook 3 to 5 minutes per side, until the fish is opaque in the center. Place the grilled fillets on individual plates, and spoon relish on top of each fillet.

SERVES 4

NEW BEDFORD

"Nowhere in all America," wrote Herman Melville in *Moby-Dick* in 1851, "will you find more patrician-like houses, parks and gardens more opulent, than in New Bedford." Today the former whaling capital of the world is a city with a past, a commercial waterfront tangle of ice houses, fish houses, docks, vacant mills, bars, and boat engine manufacturers, with the smell of brine in the air.

Old-timers remember Carmine "Fish" Romano, the mafia crime boss who ruled Fulton Fish Market in New York until he was convicted of racketeering in 1982. With a 12-year federal sentence, Romano was banned for life from working at Fulton, which had been run by mobsters since the 1920s. Romano had also run a bar off the South pier at Fulton, where he'd take care of the New Bedford guys who delivered fish; upstairs he rented space to Local 359 (United Seafood Commercial Workers Union). A member of the Genovese crime family (a codfather!), Romano got early parole and married a gal from New Bedford, moved up to Massachusetts, and resumed working in the fish business until his death in 2011.

For years the New Bedford docks flowed with "shack," a waterfront tradition of cash paid for fish and scallops, its name derived from the wooden shacks that fish buyers set up on the docks decades ago. Shacks and nightriders — boats that came in at night and were met by a guy with a truck and cash — are a thing of the past, and the old time fish buyers with names like Breezy and Doggy are gone. Urban renewal hasn't been kind to the city — a highway bisects the historic district from the waterfront, which pumps money into the local economy but is walled off from it.

Crystal Ice — an ice vending machine under the bridge — runs 24 hours a day to service the scalloping. The National Club, a seedy fisherman's bar where I wrote my first newspaper story in 1978 (about a snake charmer named Tina who performed with a live boa on Friday nights), still exists, though you won't find it on Facebook or even in the phone book.

Though the scallop industry is booming, those who work here are cautious — they remember when the commercial fishing industry collapsed in the '80s. There's much talk of the Magnuson-Stevens Act (originally the Fisheries Conservation and Management Act of 1976), which governs fisheries management in the United States, a direct response to overfishing and the desire to protect ecosystems.

Congress "needs to hit the reset button," the mayor of New Bedford bellowed recently, attempting to rebuild a fishing industry that's lost 50 shoreside businesses since 2004 and 300 jobs since 2010. It's been more than 20 years since 200 Chinese laborers were smuggled into the Whaling City aboard *Lady Diane* and loaded into a U-Haul, but New Bedford remains a gritty place, challenged by drugs, poverty, and unemployment.

"The docks are insulated from landlubbers by a dark seam of willful sinning which runs through the place, by a lingua franca as salty as it is blue, by arcane rites and rituals no less impenetrable to outsiders than the Mafia's code and by the watermen's fidelity to the sea — its gods, traditions and myths — and to each other," observed Rory Nugent, author of *Down at the Docks.* "Around the world, working waterfronts form a confederation of seafaring tribes connected by a common heritage and workplace . . . . Pecking orders and systems of justice vary from port to port, each derived from hundreds of years of precedent, with the mechanics known to battle-scarred veterans and few others."

"The waterfront is a tangle of contradictions," wrote the *New Bedford Standard Times.* "It has been maligned and celebrated, it has brought the city riches and addictions, huge triumphs and massive problems. It still offers out the great promise of fishing anyone brave enough and smart enough to go and catch fish can will himself to a brighter future. The rewards from life on the harbor can be great — and the risks enormous."

The days are long gone when the safety equipment on a New Bedford trawler consisted of some life jackets, a few flares, and eight strings of rosary beads, but according to the Bureau of Labor Statistics, commercial fishing remains the most dangerous occupation in the United States.

Sole en Papillote

My go-to meal in the winter when I'm in a hurry, this is a no-fuss way
to cook fish. The sole can be prepared several hours ahead of time and kept
in the refrigerator until ready to cook. You can substitute another white, lean,
firm fish such as pollock, haddock, halibut, or flounder.

4	(4- to 6-ounce) sole fillets
½	teaspoon kosher salt
½	teaspoon black pepper
2	cloves garlic, minced
2	tablespoons extra-virgin olive oil
2	tablespoons basil chiffonade (see page 53)
1	pint grape tomatoes, halved
2	scallions, thinly sliced

1. Preheat the oven to 400°F (200°C).

2. Cut four 12-inch square sheets of parchment paper and fold each one in half,
then open it up and put a fillet on one side of each crease line. Salt and pepper
each fillet, sprinkle with garlic and olive oil, and lay the basil, tomatoes, and
scallions evenly across the top of each fillet. Close the parchment and crease
the edges of the three sides together in a narrow fold to seal.

3. Lay the packets on a large baking sheet and bake 10 to 15 minutes, or until
the packets are puffed and golden. To serve, place the packets on plates and
slice open with a knife, taking care not to get burned by the steam.

SERVES 4

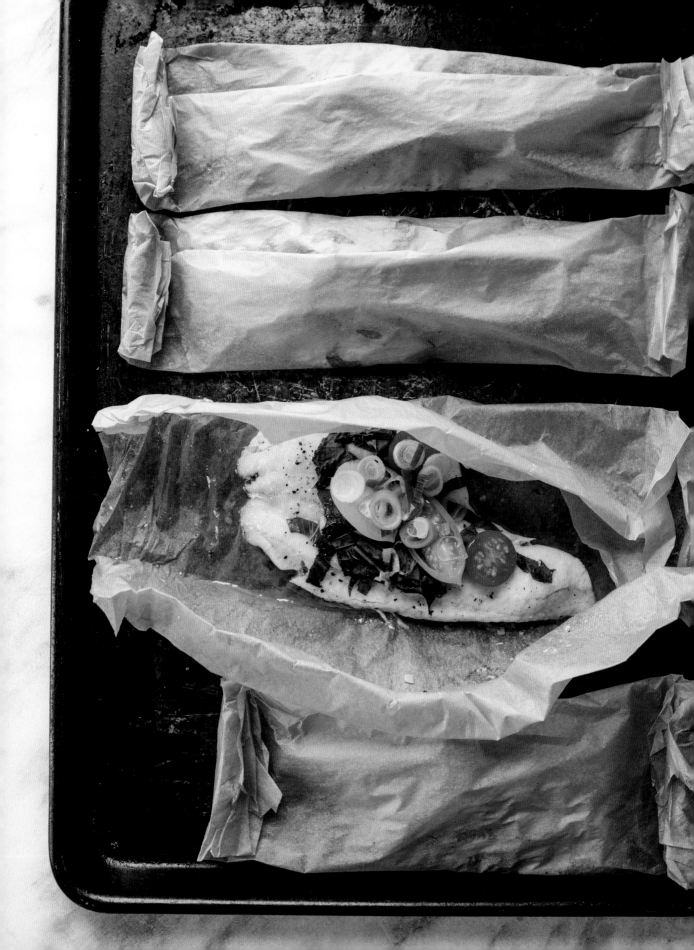

Wasabi Grilled Tuna *with* Couscous *and* Avocado-Mango Salsa

The best way to prepare fresh tuna? The simplest way. Let it shine. If you can't find a mango for the salsa, substitute papaya or a summer peach.

4	(6- to 8-ounce) excellent-quality tuna steaks, 1 inch thick
1	tablespoon extra-virgin olive oil
	Salt and freshly ground pepper

WASABI SAUCE

¼	cup soy sauce
3	tablespoons rice wine vinegar
1	tablespoon honey
2	teaspoons sesame oil
1½	teaspoons wasabi powder

COUSCOUS

1	cup whole-wheat Israeli couscous
1¼	cups vegetable stock or water
1	cup cherry tomatoes, halved
1	scallion, finely chopped
1	tablespoon small capers
2	tablespoons chopped basil leaves
2	garlic cloves, minced
¼	cup extra-virgin olive oil
2½	tablespoons lemon juice
	Salt and freshly ground black pepper

AVOCADO-MANGO SALSA

1	ripe Hass avocado, peeled, pitted, and diced
1	ripe mango, peeled, pitted, and diced
½	small red onion, diced
2	tablespoons cilantro, chopped
½	teaspoon red Scotch bonnet chile, or your favorite chile, seeded and diced
	Juice of 1 lime
2	teaspoons extra-virgin olive oil
¼	teaspoon salt

1. Prepare a hot fire in a gas or charcoal grill.

2. Lightly rub the tuna steaks on both sides with oil and season with salt and pepper. Grill the tuna 1 to 1½ minutes per side, depending on preference; the center should be red or pink.

3. To prepare the sauce: Whisk the soy sauce, rice wine vinegar, honey, sesame oil, and wasabi powder in a small bowl. Taste, and add more wasabi if desired.

4. To prepare the salsa: Put the avocado, mango, onion, cilantro, chile, lime juice, olive oil, and salt in a bowl and toss gently to combine. Allow flavors to blend for half an hour.

5. To prepare the couscous: Toast half the couscous in a dry skillet over medium heat. Bring the stock to a boil in a medium pot, stir in all the couscous, cover, and simmer over low heat until the stock is absorbed and the couscous is tender, about 10 minutes. Remove from heat, fluff with a fork, and set aside.

6. Put the tomatoes, scallion, capers, and basil in a serving bowl. Stir in the couscous. Whisk the garlic, olive oil, lemon juice, and salt and pepper to taste in a small bowl and pour over the warm couscous, tossing to combine. Serve warm or cool, with a dollop of salsa on the side and small bowls for the dipping sauce.

SERVES 4

HEMINGWAY'S NANTUCKET

Although the mention of Hemingway conjures up deep-sea fishing off the Florida Keys, his first saltwater fishing experience was on Nantucket Sound. Growing up in a Chicago suburb, his mother (the granddaughter of a sea captain) had summered on Nantucket and wanted her children to experience the same pleasures and know their nautical heritage. It sounds like she also wanted a break, because she came up with an idea: each summer she'd leave her physician husband and five of her six kids, and head to Nantucket for a month. The year that each child turned 11, he or she would accompany her to Nantucket. Ernest's turn came in 1910; it was the first time he sailed on salt water or marine-fished (he caught a bass and mackerel). He also met an old fisherman who told him a yarn about struggling to catch a swordfish.

Seared Gloucester Bluefin Tuna with Wasabi Dipping Sauce

Fishing charters are an emerging business that's booming. Guys with a boat will take you out for a day of deep-sea fishing. Gloucester is one such place, ideal for bluefin tuna fishing expeditions due to its close proximity to fishing grounds to the north, east, and south, including Jeffries Ledge, Tillies Bank, Stellwagen Bank, and Cape Cod Bay. As one charter guy told me, "A successful day is a nice 200-pounder in the boat."

⅓   **cup soy sauce**

¼   **cup rice wine vinegar**

½   **teaspoon finely minced ginger**

2   **teaspoons wasabi powder**

1   **scallion, thinly sliced**

4   **tuna steaks, approximately ½ pound each**

     Sesame oil for brushing

2   **tablespoons sesame seeds**

1. Mix the soy sauce, vinegar, ginger, wasabi powder, and scallion in a small bowl.

2. Lay out the tuna steaks, brush both sides with sesame oil, and then coat both sides with the sesame seeds.

3. Heat a cast-iron skillet over medium-high heat until it is screaming hot. Sear each tuna steak for 2 minutes per side. Set on a cutting board for 5 minutes, then slice and serve with the dipping sauce.

SERVES 4

"I have made up my mind now to be a sailor's wife,
To have a purse full of money and a very easy life,

For a clever sailor husband is so seldom at his home,
That his wife can spend the dollars with a will that's all her own,

Then I'll haste to wed a sailor, and send him off to sea,
For a life of independence is the pleasant life for me...."

— "THE NANTUCKET GIRL'S SONG," C. 1855

OH, BUOY!

WHEN WHOLE FOODS removed bluefin tuna from its sustainable list, the guys with a rod, reel, and boat in Gloucester weren't happy. People have been deep-sea fishing off Cape Ann since 1626. Is tuna overfished? It's complicated.

Sleek and muscular, bluefin tuna are the fastest swimmers in the ocean. They're a thrill to catch, prized by sport fishermen. They're also a thrill to watch — a school of them might have a spectacular feeding frenzy off the point at Provincetown, then flick their tails and reappear hundreds of feet away in seconds.

World travelers, they cross oceans chasing herring, mackerel, eels, squid, and other invertebrates, returning yearly to the Gulf of Mexico or the Mediterranean to spawn.

They are also delicious (especially from midsummer to October, when the fish have had a chance to feast on their favorite bait), which is part of the problem; though the Japanese have been enjoying sushi for over a hundred years, the market exploded when Americans caught the bug in the '70s. It's a lucrative catch.

Because tuna is an international traveler, it is managed (many say mismanaged) by the International Commission on Conservation of Atlantic Tunas, and since 2009 has been the poster child for extinction. When President Obama traveled to Japan in 2014 and ("how could he?") ate raw tuna in sushi, it made world news.

However, back in New England, tuna caught in the cold waters off Massachusetts,

New Hampshire, and Maine actually gets decent marks for sustainability. Although large nets (called purse seines) are used to a limited extent in domestic bluefin fisheries, most tuna in New England are caught with rod and reel or harpoon. Purse seines aren't necessarily bad — they were used by Stone Age societies to catch fish; what's problematic is that these massive nets catch other species, along with juvenile tuna, which are then held in pens and removed from their ecosystem.

There is little chance that we'll ever drive a species to extinction using fishing rods or harpoons. The best defense? Ask where your fish comes from, and insist on U.S.-caught fish. Otherwise, just say no.

Simplest Swordfish

Fresh swordfish is tender and meaty, more akin to scallops than to other fin fish. Whether you've reeled it in yourself, found a great fish market on Cuttyhunk, or been to the Mount Hope Farmers' Market in Bristol, Rhode Island, swordfish is a late summer treat in New England. August (when the water's warm) is the season for Atlantic harpoon swordfish — it's when fishermen go out in small boats and harpoon the swordfish by hand. The catch is brought in daily and sold at local fish markets, delivering the fish from the sea to your plate in less than 24 hours. Low in fat and high in the omega-3 fatty acids that are good for your heart and eyes, swordfish grills up as easily as a steak.

> 2   **swordfish steaks, skin on**
> **Olive oil**
> **Salt and freshly ground black pepper**
> **Lemon wedges, for garnish**

1. Prepare a medium-high fire in a gas or charcoal grill.

2. Coat the steaks with the oil and season with salt and pepper. Grill the steaks for 4 to 6 minutes per side (the time will depend on the thickness of the steaks), taking care not to overcook. Serve with lemon wedges.

SERVES 2

Block Island Swordfish with Lemon-Fennel Butter

Around 25,000 years ago a huge sheet of ice a mile thick plunged down over New England, pushing a chunk of the mainland offshore, creating a string of islands that parallel the coast from New York to Massachusetts — Long Island, Fishers Island, Block Island, the Elizabeth Islands, Martha's Vineyard, and Nantucket. Where Block Island sits, 13 miles offshore, was once frozen tundra, extending 70 miles farther out to sea. Indeed, fishermen using big nets to catch bottom fish off Block Island still occasionally pull up mammoth and mastodon molars. As the glacier retreated, all that melting ice created a great runoff of meltwater, forging deep rivulets that today are essentially underwater canyons off Block Island. Swordfishing is great in these deep waters.

1	tablespoon olive oil
5	tablespoons butter
4	(6-ounce) swordfish steaks
2	teaspoons fennel seeds
1	teaspoon lemon zest

1. Heat the oil with 1 tablespoon of the butter in a skillet over medium-high heat. Add the swordfish and fry until just cooked through, about 6 minutes per side, depending on thickness.

2. While the fish is cooking, heat the remaining 4 tablespoons of butter in a small saucepan over medium heat with the fennel seeds and lemon zest. Cook 6 minutes, or just until the butter is on the verge of browning. Transfer the swordfish to plates, and spoon the sauce on top.

SERVES 4

> COOKING FISH
>
> The general rule of thumb is to grill or broil fish 10 minutes per inch of thickness.

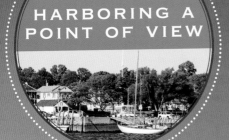

HARBORING A POINT OF VIEW

Cruising on a boat is a state of mind. When I was growing up, we went on overnight trips several times a summer, sailing to islands and harbors with great spots to fish, swim, explore, beachcomb, and gunkhole. Some were tony or honky-tonk summer resort towns, others were remote islands with an abandoned cottage or two. There are many beautiful harbors in New England — here are favorites from readers polled by *Yachting* magazine.

BOOTHBAY HARBOR, MAINE

The "boating center of New England," with several marinas within walking distance of the village, and Monhegan Island a short sail away.

EDGARTOWN, MASSACHUSETTS

Elegant, patrician, with magnificent anchorages, sandy beaches, and picturesque whaling-captain's homes.

ESSEX, CONNECTICUT

A charming spot with yacht clubs, gentrified marinas, and interesting cruising grounds at the mouth of the Connecticut River.

MYSTIC, CONNECTICUT

A stop for mariners, where you might just see the *Charles W. Morgan* (last surviving whaling ship) under sail. An old shipbuilding town that hasn't lost its saltiness despite the crowds and busy interstate nearby.

NEWPORT, RHODE ISLAND

A jaw-dropping harbor for boat watching, where you can sail a 12-Meter, learn about America's Cup history, watch world-class regattas, and spiff up for the New York Yacht Club when it comes to town.

PORTLAND, MAINE

A cool city by the sea and great boating location for boats of all sizes. Plus, there are more than 200 islands in Casco Bay.

It's a great list, and I know I can only get into trouble by adding a few other favorites: the Thimble Islands, Galilee, Point Judith Pond, Dutch Harbor, Kickamuit River, Third Beach, Tarpaulin Cove, Marion, Onset, Cuttyhunk, Boston Harbor, Isles of Shoals, Matinicus Harbor, Brooklin, Castine, Stonington, Mackerel Cove, Frenchboro, Southwest Harbor . . . or as a friend from Maine wrote: "sticking with Penobscot Bay, which is the best of the coast of New England, for sure — North Haven and the Fox Island Thoroughfare, Stonington, Camden, Rockland, Belfast, and Dark Harbor. Venturing a little further downeast, Winter Harbor. To the south and west of the Maine coast, where I grew up, I'll leave to someone else. Once you get to New Hampshire, it's pretty much over." Spoken like a true Downeaster.

Grilled Swordfish *with* Pasta *and* Summer Tomato Sauce

This no-cook tomato sauce is ideal on hot summer nights.

SAUCE

4	large ripe tomatoes, chopped (about 3 cups)
1	tablespoon basil chiffonade (see page 53)
1	garlic clove, finely minced
2	tablespoons extra-virgin olive oil
1	tablespoon balsamic vinegar
2	tablespoons chopped Kalamata olives
1	tablespoon drained capers
¼	teaspoon red pepper flakes
¼	teaspoon salt
½	pound thin spaghetti
4	swordfish steaks, approximately ½ pound each
	Grapeseed oil
	Salt and freshly ground black pepper
½	cup grated Parmesan or Romano cheese, or a combination

1. To prepare the sauce: Combine the tomatoes, basil, garlic, oil, vinegar, olives, capers, pepper flakes, and salt in a bowl, and set aside for at least 30 minutes to meld the flavors.

2. Prepare a medium-high fire in a gas or charcoal grill.

3. Bring a large pot of salted water to a boil and cook the pasta until al dente, 8 to 10 minutes.

4. Meanwhile, brush both sides of the swordfish steaks with oil and season with salt and pepper. Grill for 3 to 4 minutes per side, or until just done.

5. Drain the pasta and toss with the tomato sauce and the Parmesan. Serve with the swordfish steaks.

SERVES 4

Deep Sea Fluke Crusted *with* Jonah Crab *and* Beurre Blanc Sauce

This comes from the chef of one of my favorite restaurants. I had it in March — it was one of the specials — and I knew the owner had gone down to the dock and picked the fish right off the boat. You could substitute another thin fish, such as sole.

BEURRE BLANC

2	shallots, minced
	Juice of 1 lemon
¼	cup white wine
2	tablespoons white wine vinegar
1	teaspoon sugar
1	teaspoon kosher salt
2	tablespoons heavy cream
1	cup (2 sticks) butter, cold and cut into small cubes
	Pinch of white pepper

1	pound fresh fluke, cut into 4 fillets
½	cup cornstarch seasoned with salt and pepper
1	cup panko breadcrumbs
4	ounces fresh Jonah crabmeat, broken up into small pieces
3	tablespoons butter, melted
	Juice of 1 lemon
2	tablespoons grapeseed oil

1. To make the beurre blanc: Put the shallots, lemon juice, wine, vinegar, sugar, and salt in a small saucepan over medium heat and simmer until the liquid is reduced to 2 to 3 tablespoons, stirring constantly. Whisk in the heavy cream and continue to reduce the liquid, about another minute. Remove the saucepan from the heat and start whisking in the cold butter a few cubes at a time until all the butter is incorporated. You will have a light and creamy butter sauce with a hint of lemon. Season with white pepper and store in an insulated thermos until ready to serve.

2. Dredge the fillets in the seasoned cornstarch.

3. In a small bowl, combine the breadcrumbs, crabmeat, butter, and lemon juice.

4. Preheat the broiler. Heat the oil in a large skillet over medium-high heat, and then sear the fillets for about 1½ minutes per side. Turn the fluke into a small casserole dish or sheet pan and top with the crab mixture. Place under the broiler until the topping is browned, about 1 minute. (If your sauté pan is oven-safe, you can skip the casserole dish and broil directly in the pan.) Transfer fillets to individual plates, spoon the beurre blanc on top, and serve immediately.

SERVES 4

"Two weeks in, the days have a comforting sameness to them: mornings at the Bluff, afternoons at the Big Cove, evenings on the piazza. We live in our bathing suits. Our feet, callused from going barefoot all day, no longer cringe on the rocky shore. Rubbed by sun, wind, and water, our city edges are wearing away. I feel as weathered as driftwood, as smoothed as sea glass. When I woke this morning I couldn't remember what day it was."

— GEORGE HOWE COLT

Fluke *with* Chipotle-Lime Quinoa

With a sweet white meat, fluke (which resembles a flounder) is a great-eating fish. It's a bottom-dwelling flatfish that's also called a summer flounder (no, it's not a fluke) because they winter off the continental shelf and come into the bays during the summer months.

QUINOA SALAD

1	cup quinoa
2	cups vegetable broth

DRESSING

	Juice of 1½ limes
¼	teaspoon lime zest
2	tablespoons olive oil
1	garlic clove, minced
⅛	teaspoon chipotle powder
¼	teaspoon salt
	Freshly ground black pepper

4	scallions, minced
¼	cup chopped fresh cilantro
⅔	cup cherry tomatoes, halved
4	fluke fillets, approximately ½ pound each
1	tablespoon olive oil
1	tablespoon butter
	Salt and freshly ground pepper
4	lemon wedges
1	ripe avocado, cut into thin slices
1	lime, cut into thin wheels, for garnish

1. To prepare the quinoa: Rinse the quinoa several times, then combine with the vegetable broth in a saucepan, cover partially, and cook over medium heat for 10 minutes. Lower the heat, cover completely, and cook an additional 10 minutes. Drain the quinoa, put in a bowl, and let cool slightly.

2. To make the dressing: Whisk together the lime juice, lime zest, oil, garlic, chipotle powder, salt, and pepper to taste. Toss with the quinoa, and then add the scallions, cilantro, and cherry tomatoes. Toss once more gently and season to taste. The quinoa is delicious at room temperature or cold.

3. Pat dry the fillets, and season with salt and pepper. Heat a heavy, flat skillet over medium-high heat. Add the oil and butter, and then place the fillets skin-side down in the skillet and cook for a few minutes, until golden, gently agitating the pan so the skin doesn't burn as it crisps. Flip carefully, and finish cooking through, about 1 minute. To serve (immediately), plate the quinoa salad and lay a fillet skin-side up atop each salad. Garnish with the lemon wedges and serve on individual plates, with avocado slices fanned over the quinoa and lime wheels on the side.

SERVES 4

FLYING FISH

When I hopped aboard my first boat delivery to the Caribbean and sailed down the Atlantic seaboard before heading out to sea, it was a seminal experience. Days and nights governed by four-hour watches, on and off, just like the eight bells in the ship's clock in my parents' living room. Waking up after my watch to find we'd drifted into a shipping lane on a moonless night and were only yards from a huge ship. My cousin asleep on deck until a flying fish leaped from a wave and slapped him on the cheek. Catching a shark. Eating it for days. Peeling away the wool layers as we sailed from a crisp October in Maine to a hot island in the tropics. Seeing shorebirds and then spotting land for the first time in two weeks. Landing in Virgin Gorda and running for ice cream and hot showers. Getting back on the boats again, and again.

After logging 25,000 sea miles delivering sailboats, I realized that — other than a well-developed sense of humor — the best attribute one can have at sea is the flexibility to respond to changing conditions. A good lesson.

Crispy Sea Bass *with* Saffron Fennel–Roasted Tomato Compote

If you know a few tricks, fish is easy to cook indoors. With this recipe, you can substitute any white fish (or salmon) with the skin on for the sea bass — you'll love the crackly-crispy taste.

COMPOTE

2	tablespoons extra-virgin olive oil
1	large fennel bulb
¼	teaspoon saffron
1	teaspoon minced garlic
¼	teaspoon salt
	Freshly ground black pepper
4	roasted peeled tomatoes, chopped (see recipe, page 293)
2	tablespoons (heaping) roughly chopped Kalamata olives
1	tablespoon small capers
1	pound sea bass fillets, skin on
	Kosher salt
1	tablespoon grapeseed oil

1. To prepare the compote: Heat the olive oil in a large skillet over medium-low heat. Thinly slice the fennel bulb and chop 1 tablespoon of the fronds; add to the skillet along with the saffron, garlic, salt, and pepper to taste. Cook until the fennel softens, about 15 minutes. Add the tomatoes, olives, and capers, and simmer for 15 minutes, stirring occasionally. Set the compote aside, keeping it warm.

2. Pat dry the fish. Line a plate with a thin layer of kosher salt and press the skin side into the plate to draw out excess moisture. Leave for 15 minutes, seasoning the top lightly with salt and pepper. Use a knife to scrape the salt and any excess water from the fish skin. Pat dry again.

3. Spread the grapeseed oil to thinly coat the bottom of a cast-iron skillet. Get the skillet good and hot over medium-high heat; it should be beginning to smoke. Sauté the fillets skin-side down, spaced apart, weighing them down gently with a lid or spatula, to prevent edges from curling up, for 1 minute. Remove the lid but continue cooking (don't mess with it) until the skin is golden brown (I know, you're going to need to peek at some point), and only the top is uncooked. The fish is now 75 percent cooked; flip and cook another minute or two. Top each fillet with the compote and serve.

SERVES 4

Herb-Roasted Whole Black Sea Bass *with* Sesame Spinach

The angler in your family has caught a gorgeous fish, and you're wondering what the heck to do with it. Simple: roast it. You could easily substitute red snapper. The prime season for black sea bass is late summer, when they migrate and spawn off New England.

4	tablespoons extra-virgin olive oil
2	garlic cloves, peeled and thinly sliced
1	small shallot, peeled and minced
1	small carrot, peeled and thinly sliced
1	celery stalk, thinly sliced
1	tomato, diced
1	(1-pound) whole black sea bass, cleaned and scaled
	Sea salt and freshly ground black pepper
3	sprigs fresh thyme plus 1 tablespoon chopped thyme
2	sprigs parsley plus 1 tablespoon chopped parsley
2	sprigs fresh rosemary
½	cup thinly sliced scallions
	Juice of 1 lemon
	Sesame Spinach (see page 292)

1. Preheat the oven to 450°F (230°C). Use olive oil to lightly brush a baking dish large enough to hold the fish without crowding.

2. Combine the garlic, shallot, carrot, celery, tomato, and 2 tablespoons of the oil in a small bowl.

3. Rinse the fish cavity under cold water and dry with paper towels. Season inside and out with salt and pepper. Stuff the cavity with the vegetable mixture and the thyme, parsley, and rosemary sprigs, and place the fish in the prepared baking dish. Sprinkle the scallions over and around the fish. Drizzle the remaining 2 tablespoons oil over the top.

4. Roast until the thickest part of the fish is just firm and the skin is blistered, about 15 minutes. Drizzle with lemon juice. Transfer the fish to a cutting board and spoon the stuffing into a bowl; discard the herb sprigs. Lift out and discard the fish bones. Transfer the fillets to plates and spoon the stuffing on top. Drizzle the fillets with pan juices and sprinkle chopped fresh thyme and parsley on top. Serve with a mound of sesame spinach.

SERVES 2

THE OUTER LANDS

Twenty-five thousand years ago, the shoreline from Long Island to Cape Cod was covered by a glacier. The earth warmed, the glacier melted, and its retreat was like a glacial bulldozer as the land 20 miles inland cooled, then heated up, then cooled, then heated up again, then cooled, pushing the outer banks out to sea and creating a nearly connected chain of peninsulas and low-lying islands (essentially glacial deposits) that still exist today. These are islands of enormous beauty: from Long Island, Fishers Island, Block Island, and the Elizabeth Islands to parts of Cape Cod, Martha's Vineyard, and Nantucket. Though spread over four states — with different politics, taxes, and school vacations — these "Outer Lands" are similar geographically, with dunes, kettle ponds, beach plums, and clams.

Toward the eastern end of this offshore archipelago, the Elizabeth Islands chain holds Buzzards Bay, an area coveted by sailors for its craggy shoreline and predictable southwesterly afternoon breeze. In the 1600s, Captain Kidd sailed here, using Naushon Island as a hideout after he raided ships in the Indian Ocean.

Halibut over Wild Rice with Spicy Cucumber Salad

Halibut can be baked, broiled, fried, grilled, or poached without complaining. Indeed, it's an iconic New England flatfish — white, firm, flaky, not too oily — and a pure protein that's been ridiculously overfished in the Gulf of Maine/Georges Bank region. Since the early 1900s, the Atlantic halibut population has been steadily declining, and, even though this is a New England cookbook, Pacific halibut is a more sustainable fish choice. I've nonetheless included a recipe for it, partly because you might bring one home from a day of fishing, and also because I remain optimistic that as more people think about sustainable food choices, commercial quotas will help manage the stock. Good substitutes include fluke, sole, flounder, wild striped bass, and cod.

½   **cup wild rice**
2   **cups water or stock**
     Pinch of salt

     Spicy Cucumber Salad (page 260)

1   **(2-pound) halibut fillet or other white fish**
1–2   **tablespoons olive oil**
     Salt and freshly ground black pepper

1. To prepare the rice: Rinse the rice and combine it with the water or stock and salt in a pot. Cover and bring to a boil over high heat, then reduce to a simmer and cook, covered, for 45 minutes, or until the rice is chewy. While the rice is cooking, prepare the cucumber salad.

2. Prepare a medium-high fire in a gas or charcoal grill. Spray the grill grate with nonstick spray. Rub both sides of the fish with oil, season with salt and pepper to taste, and cook over medium-high heat for 5 minutes per side, covered, or until the fish is opaque and flakes easily.

SERVES 4

> **NOTE**
> Halibut dries out fast because it has little oil, so if grilling or baking it, marinate it or brush with olive oil.

SEA SALT

In colonial times, salt was the key to preserving fish, meat, vegetables, and cheeses, and exporting salt cod was big business. While early colonists made their own sea salt, the New England fishing industry really took off at the end of the 1600s when they started importing salt from the British Caribbean.

In 1767, English Parliament passed the Townsend Revenues Act, taxing a number of goods, including salt, just one in a series of events that led to the American Revolution. After the Revolution broke out, British ships patrolled Vineyard Sound and Buzzards Bay, enforcing the embargo on all shipping, and seizing boats, cargo, and crew. Cape Codders couldn't get salt. Increasingly concerned about a nascent nation without salt, the Continental Congress in 1776 jump-started the industry, telling colonists they'd pay one-third of a dollar for every 80 pounds of salt (requiring 350 gallons of seawater) they could produce, and printed instructions to get people going on it.

Cape Cod and the Islands have an abundance of seawater, sun, and wind, and by 1800, windmills were a familiar site along the shore, pumping seawater into evaporation vats. (In many seaside towns, the local economy centered on whaling, shipbuilding, fishing, and salt making.) On the Vineyard, the first known salt works was running by 1778, and by 1807, it was the island's second largest industry. Salt reached $8/bushel in 1783 and was shipped to the West Indies, Europe, and down south in small coastal schooners. On Cape Cod, there were 658 salt works by 1837, producing 26,000 tons of sea salt each year. Falmouth alone had 1.8 million square feet of salt works; it was a short sail from Falmouth to New Bedford and Nantucket, where salt was in high demand for whalers preparing for long voyages.

The beginning of the end (one might say they hit rock bottom) was the opening of the Erie Canal and discovery of salt deposits around Syracuse, making inland salt suddenly available cheap. The death knell was canning and refrigeration, diminishing the need for salt-preserved foods. The last salt works of Falmouth was sold to developers in 1871 who built a summer resort hotel on the site — marking the beginning of a new industry (the railroad extended to Woods Hole in 1872).

Today, from Cape Cod to the Canadian border, artisanal salt makers are reviving the art of making sea salt, bringing the briny taste of the sea to the table with 100 percent natural salt with no additives or anticaking agents. You'll find Kenyon's in Rhode Island; Cape Cod Saltworks, Wellfleet Sea Salt, and others on the Cape; Martha's Vineyard Sea Salt and Ambrosia Nantucket on the islands and Maine Sea Salt (near Machias) and Quoddy Mist in Maine (from the Bay of Fundy), to name just a few. Many have enhanced flavors, like Dulse Seaweed Maine Sea Salt, or Nantucket's Smoked Black Lava salt.

"THERE MUST BE SOMETHING STRANGELY SACRED IN SALT. IT IS IN OUR TEARS AND IN THE SEA." — KAHLIL GIBRAN

3

THINGS WITH

SHELLS

LET'S TALK ABOUT LOVE. A bivalve revered since ancient Rome (when oysters were imported from England, put in saltwater pools, and fattened up with wine), oysters have been eaten on the coast for at least a thousand years, evidenced by mounds of discarded oyster shells (called middens) that line the banks of Maine's Damariscotta River and other points along the shore. As the Colonies developed, oyster bars sprang up; when Boston's Union Oyster House opened in 1826, Daniel Webster came daily, drinking a tall brandy and water and often eating six plates of a half-dozen oysters.

Long considered an aphrodisiac because of its high zinc content (in the Middle Ages oysters were thought to "exciteth Venus," while in the 18th century Italian lover Casanova purportedly ate 50 for breakfast), the sensual oyster has some scientific merit: scientists in 2005 discovered that oysters are rich in a rare amino acid that triggers sex hormones.

What is it about this creature that outlived the dinosaurs, and can be confidently dished up raw in a multitude of situations — from a fancy wedding to a beachy clambake or a roadside fish shack? That can be stuffed, boiled, pickled, smoked, baked, stewed, steamed, grilled, canned, iced, or fried? Ardent environmentalists, oysters clean our waters with constant filtration, moving up to five gallons an hour across their gills. Rich in minerals, high in protein, they are a nutritional juggernaut. Farmed or wild, they connect us to our environment, our earth, our seas.

Oysters used to be cheap — one of the earliest fast foods, they were sold on city streets, where oyster houses and oyster bars proliferated. In the 1800s, New Haven became an oyster center with its vast oyster population and a railroad system. Oyster farming thrived in the Ocean State, too, where in the early 1900s, oyster growers leased 30 percent of the bottom of Narragansett Bay for underwater farms. Today, from New Haven, Connecticut, to Mattapoisett, Massachusetts, you can still find Greek Revival–style oystermen's houses on the shore, architectural gems that were practical: a place to live above and store their catch below. Peaking in the early 1900s, New England oyster production began a slow, steady decline as oyster beds and wild oyster reefs became nearly extinct, wiped out by overfishing,

OYSTERS ARE A MICROCOSM OF OUR RELATIONSHIP WITH NATURE AND SEA.

a devastating hurricane in 1938, urban sprawl, failed septic systems, pollution, and bacteria that thrive on pollution. By the 1970s, many people wouldn't eat raw oysters. As Woody Allen said, "I want my food dead. Not sick. Not wounded. Dead."

There's been a resurgence, beginning in the late 1990s, that wouldn't have been possible without the Clean Water Act and other efforts to restore the ecosystem of our bays, waterways, rivers, and ocean. But cleaner waters also dovetailed with the foodie revolution, leading to an oyster renaissance where East Coast oyster production has doubled in the last five years. While the big guns who supply the bulk of commercial oysters (sold by the gallon for fries and stews) are located in Virginia and Louisiana, New England raw oysters grown by small-craft farmers have achieved cachet status. The days of being served a plate of Malpèques (a popular oyster from Canada) are over — people want Duxbury, Cuttyhunk, or hundreds of other named oysters.

Barriers to entry are few. Many New England oyster growers are young, according to Perry Raso, who started the Matunuck Oyster Farm at age 26. He's typical: a native New Yorker, he came to Rhode Island as a boy when his dad retired as a cop and decided to try his hand at fishing. That didn't last, but 12-year-old Raso was already making money selling the clams he'd dug, and caught the bug. Today, he owns the popular Matunuck Oyster Bar, where New Yorkers drive three hours to have their Matunuck oysters brought from pond to plate. It's a lifestyle as much as a job — good thing, since you may think it's the best job in the world on a hot summer day, but not so much in the middle of January when you're standing hip-deep in finger-numbing water.

Or consider Walrus and Carpenter Oysters in Kingston, started by a Yale forestry school grad who wanted to work on the water with sustainable aquaculture. Walrus and Carpenter supplies 25 Rhode Island restaurants, shucks at parties, and sells to individuals. In December, the owner drives weekly to New York and Brooklyn (he's no fool), offering free beer, shucking lessons, and oyster samples to folks who

come to the evening pick-ups, mostly because he wants to connect individuals to their farmer. In 1996, Rhode Island had six oyster farms; today there are thirty.

Oysters reproduce by spawning when water temperature reaches about 68°F (38°C). They grow faster in warmer waters (sometimes ready to harvest in a year), whereas up north it takes longer — two or three years for an oyster to mature — and oyster connoisseurs believe the slower-growing, cold-water varietals have more flavor. Considering that all oystermen start with the same seed, it's amazing how different oysters can taste; artisanal growers rely on marketing these distinctive flavors to enthusiasts who appreciate (or at least are curious about) the special characteristics of geography and climate of a place — what the wine industry calls *terroir*. Flavor profiles are determined by many factors — what plankton the shellfish eat, water salinity, time of year, tidal flow, temperature, and the water's nutrient-richness.

Consider Maine, where the cold water slows the oyster's growth (taking up to four years to mature), yielding a firm plump meat, thick shell (easier for shucking), and distinct brininess. On the Damariscotta River alone, farms closer to the mouth (where the water is saltier) produce a brinier oyster with a salty finish, while farms upriver yield mellower, sweeter oysters. Barbara Scully, a former marine biologist who founded the Glidden Point Oyster Sea Farm, is among the few farmers who don a wetsuit and dive for oysters. (Her kids chip barnacles off the shells when she surfaces with them.) Sean Rembold, chef at the Wythe Hotel in Williamsburg, Virginia, says the extra care taken with Glidden Point oysters produces a buttery briny taste that is "majestic."

New Hampshire, with the shortest shoreline of any coastal state, has a handful of oyster farmers in the Great Bay Estuary, and The Nature Conservancy, recognizing what an ecological linchpin oysters are, teamed up with the University of New Hampshire to restore oyster reefs in the Great Bay and provide the natural filtration necessary to maintain healthy eelgrass beds.

In Massachusetts, Wellfleet oysters have a lineage dating back to days when whaling and oystering drove the local economy, and lovers of Wellfleet oysters insist that the shallow bays turn out lighter-bodied oysters than you'll find off the islands

of Nantucket Sound. Just west, Island Creek Oysters in Duxbury benefit from cold, hyper-salty water and a tidal system that flushes out the rivers 80 percent with each tide — producing a crisp, firm, super-briny oyster. Island Creek, which started growing clams in 1990 (when there were no other aquaculturists in the area), is a collective of 15 farmers who work together selling their designer oysters to French Laundry, Per Se, and other renowned restaurants where raw oysters have become a menu mainstay. In Connecticut, Thimble Island oysters started the first community-supported fishery (CSF) in Connecticut and Long Island Sound history, creating a way for local residents to support their efforts to restore the shoreline's ocean ecosystem and enjoy their oysters.

A small percentage of oysters sold today are caught wild; oysters are usually farmed in nets and trays suspended in the water, requiring little or no chemicals. Given that oysters filter plankton and improve local water conditions, oyster farming is considered a good environmental choice. With a seemingly insatiable half-shell market, there is constant demand and decent profits for farmers who can succeed in this burgeoning business. But there remain uncertainties, notably the impact of climate change, warming waters, and increased ocean acidity, not to mention property owners and boaters who don't always like sharing narrow estuaries with pens and tenders.

Regardless, oysters are good for the earth and for you, and many committed farmers are working to ensure that their environment is sustainable. One of New England's success stories, oysters are a microcosm of our relationship with nature and the sea, seen through the prism of nature's pearl. This chapter is about oysters and other mollusks that we harvest from estuaries and coastal waters.

Naked Oysters with Citrus Mignonette

Raw and unadulterated — that's confidence food. Freshly shucked?
It doesn't get much saltier. If you shuck your own, you'll really impress someone.
It's a beachside ritual, Zen in spirit.

CITRUS MIGNONETTE

- 2 tablespoons good champagne or red wine vinegar
- 2 tablespoons minced shallots
- Freshly ground black pepper

- 8 fresh oysters, cleaned and halved
- Lemon wedges
- Hot sauce (optional)

1. To make the mignonette: In a small bowl, combine the vinegar, shallots, and pepper to taste. Let this mixture rest for 20 minutes to soften the shallots.

2. Arrange the oysters on a bed of ice. Place the mignonette in a ramekin in the center with a small spoon, so guests can drizzle the dressing onto their chilled raw oysters. Some people prefer a squirt of lemon and a shake of hot sauce.

SERVES 2-4 AS AN APPETIZER

"Food and meals are influenced by the context in which they are consumed. Slurping oysters in Paris at Le Dâme with a tower of seafood in front of you makes you feel like a king. Eating them on the dock in Menemsha as the sun sets over the ocean in the autumn, when the crowds have gone and your sweater still smells musty from storage, makes you feel calm and blessed."

— CHRIS FISCHER, CHEF OF THE BEACH PLUM RESTAURANT, MARTHA'S VINEYARD

IN THE RAW

Having a party? Consider a raw bar. Lay some oysters from East Dennis, Duxbury, and Wellfleet on a bed of shaved ice and decorate with seaweed. You'll need a few good shuckers to make your raw-bar party really rock. Stock the bar with hot sauces, fresh horseradish, mignonettes (pages 144 and 150), and lemons. Oysters pair well with microbrewed beers and artisanal rum.

SHUCKING OYSTERS

Resembling a hair comb Marie Antoinette might have worn, oysters can be intimidating, but needn't be. First, don't panic. Shucking them is not that hard. And you will so impress your guests if you shuck them at the last minute (plus, the oysters won't dry out). At an annual summer family party in Duxbury, we shuck raw Duxbury oysters outdoors on a picnic table near the bonfire, with only a bottle of hot sauce and a few lemon wedges, while everyone stands around, greedily slurping them on the spot.

You'll need heavy rubber or mesh gloves, an oyster knife (the best are slightly curved on top), and an old kitchen towel or rag. If you can't find an oyster knife, use a flat-head screwdriver — you need something thin enough to slip between the shells but strong enough to pry it open.

Keep the oysters cold and buy them the day you shuck them, if possible. Don't suffocate them in plastic. Discard any oysters that have opened. Rinse them under cold water with a stiff brush to remove dirt, and with a knife, remove any barnacles that might get in the way of slurping.

Place the oyster in the center of the kitchen towel and hold it down with your hand. Insert the tip of the knife into the hinge (it's indented, where the shells come together, and no two hinges are alike) and jimmy open the shell.

Run the knife around the edge to loosen the oyster, taking care not to dump the liquid — it's delicious, and contributes to the salty brininess of the experience. Discard the top shell, and cut the muscle holding the meat to the shell so you can slurp it. (Don't rinse the oyster.) Serve on a bed of ice if you aren't eating immediately.

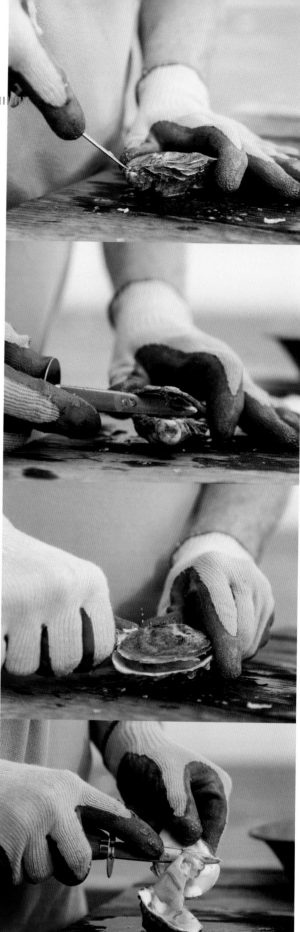

Grilled Oysters *with* Habanero Sauce

Don't like shucking? Put oysters on the grill whole. Voilà. Easy. This versatile Caribbean-style sauce also goes well with clam cakes, fried clams, and other seafood appetizers.

HABANERO SAUCE

1	ripe papaya, peeled, seeded, and chopped
½	cup white vinegar
1	shallot, coarsely chopped
1	habanero or Scotch Bonnet chile, seeded
½	teaspoon fresh minced ginger
½	teaspoon fresh minced garlic
½	teaspoon curry powder
	Juice of ½ lime
1	tablespoon agave
16	oysters

1. To make the sauce: Blend the papaya, vinegar, shallot, chile, ginger, garlic, curry, lime juice, and agave in a food processor until smooth, about 2 minutes. Transfer to saucepan and cook over low heat for 10 minutes.

2. Prepare a medium-hot fire in a gas or charcoal grill. Place the oysters on the grill, cup-side down, and close the lid, cooking them until they start to peek open, anywhere from 3 to 10 minutes. (Discard any oysters that don't open.) Remove the oysters from the grill immediately and pry them open the rest of the way with a fork, taking care not to spill the precious juices. Serve in the shell, with the habanero sauce for dipping or spooning.

SERVES 4

OYSTERS: PRACTICAL MATTERS

SIZE, SHAPE, AND TASTE vary, depending on location and the characteristics of local waters, which are as key as grape regions are to varietals. As a rule, the deeper the water, the healthier the oyster. (Oysters found in deep water also have less brittle shells.)

THE FLAVOR OF AN OYSTER also depends on the season — in the fall and winter, you'll find sweeter oysters because the oysters have stored up sugar to insulate themselves from the cold; summer oysters are brinier and crisper.

WHEN IS AN OYSTER FRESH? When you stick your nose up to the shell, you should smell the sea air. If you get that low-tide, eelgrass-marsh, slightly sour smell, forget it.

WITH SAMPLING OYSTERS, trying ten different types is like sipping ten different wines; each flavor is impacted by the previous. Rather, try a few of each — say, three of one variety, three of another — to better savor the distinctions.

I don't know who came up with the idea of shooting an oyster and swallowing it whole, but not only is it slimy and sort of gross, you miss the true taste of the oyster: the plump briny meat with its bright flavor, slightly metallic sea taste, and modestly sweet finish. When William Thackeray toured America in 1852, he was served oysters at the Tremont House Hotel in Boston. When asked how he should eat this "animal," he was told to swallow it whole. He was so disgusted that he blurted out, "I feel as if I'd swallowed a live baby."

THE R RULE about only eating oysters in months with an "r" harkens back to days before refrigeration and is largely retired because oysters are often farmed and carefully monitored. However, with raw wild oysters, stick to fall, winter, and spring; they spawn in summer, which makes them watery, soft, and less flavorful. Some connoisseurs still believe oysters taste best in the winter, when the beds are the coldest (though with farmed oysters you can eat them anytime).

WHAT ABOUT THE PEARL? Don't worry — you won't break a tooth; pearl oysters are harvested strictly for pearl production. To find one spontaneously in the wild is extremely rare — your chances are 1 in 10,000. But here's hoping!

Matunuck Oysters WITH Ginger Mignonette

Matunuck, Rhode Island, is near the fishing village of Jerusalem, which is across the harbor from the road to Galilee. (You can't make this stuff up.) Today, a half dozen oysters is considered a serving (in Europe, it's a dozen) — a far cry from appetites in colonial times, when men would eat oysters by the gross (roughly 10 bushels each year per capita). This recipe is from the Weekapaug Inn nearby, where they drizzle a ginger-shallot vinaigrette over oysters, and finish them with fresh shaved cucumber and a mango granita.

GINGER MIGNONETTE

1	tablespoon minced shallot
½	cup rice vinegar
½	teaspoon minced ginger
	Freshly ground black pepper, 6–8 turns
¼	teaspoon salt

MINT-PASSIONFRUIT GRANITA

½	cup passionfruit or mango purée (1 mango makes ¾ cup mango purée)
1	cup water
1	teaspoon honey
¼	cup orange juice
2	tablespoons lime juice
1	sprig mint
24	raw oysters, shucked and in the shell
1	(3-inch) piece English cucumber, finely diced
	Tabasco sauce (optional)

1. To make the mignonette: Combine the shallots, vinegar, ginger, pepper, and salt in a bowl and stir to incorporate. Refrigerate until ready to use.

2. To make the granita: Combine the fruit purée, water, honey, orange juice, and lime juice in a saucepan and bring to a simmer over medium heat. Remove from the heat, add the mint, and steep off the heat for 10 minutes. Strain through a chinoise or fine-meshed sieve and freeze until solid, about 2 hours. Remove the mixture from the freezer and scrape with a fork, then return to the freezer until ready to serve.

3. To assemble the oysters: Dress each oyster with a teaspoon of the mignonette, making sure to include some of the shallot and ginger pieces. Top each oyster with eight or so finely diced pieces of fresh cucumber. Right before eating, finish each oyster with ⅛ teaspoon of the granita. Serve on ice immediately with Tabasco sauce if desired.

SERVES 4-6 AS AN APPETIZER

UNDERWATER FARMERS

South Norwalk, Connecticut, is local and salty, gritty and coming on, with tequila bars around the corner from old oyster joints and places like Hillard Bloom Shellfish Company, which has been farming and harvesting oysters since oyster rights to underwater acreage were granted by the King of England before the American Revolution. The late 1800s were the heyday of Connecticut oystering; people still talk about how oysters virtually disappeared from Long Island Sound from 1955 to 1965 — they don't know whether it was pollution, hurricanes, or a natural cycle. Selling clams to Campbell's Soup kept Hillard's in business, and since the 1970s they've seen a boon, with Hillard Bloom now owning about 11,000 underwater acres from Greenwich to Branford. "We're underwater farmers," said owner Leslie Miklovich simply, looking out at a few of her 11 boats tied up to the docks, including *Bivalve*, a wooden oyster boat from the early 1800s.

Cornmeal-Crusted Oysters WITH Old Bay Sauce

New Orleans–style fried oysters make an excellent appetizer. Put them in a bun, and you're approaching a Po' Boy.

OLD BAY SAUCE

¼	cup mayonnaise
¼	cup sour cream
½	teaspoon Old Bay seasoning
3	teaspoons minced scallions
1	teaspoon smoked paprika
	Jump Up and Kiss Me Original or other hot sauce

½	cup cornmeal
1	teaspoon kosher salt
1	teaspoon black pepper
1	teaspoon cayenne pepper
12	raw oysters, shucked
¼	cup maple syrup
	Canola oil

1. To make the sauce: In a small bowl, combine the mayonnaise, sour cream, Old Bay, scallions, paprika, and hot sauce.

2. Combine the cornmeal, salt, pepper, and cayenne in a medium bowl and stir to blend. Drain the oysters and pat dry with paper towels. Dip the oysters in the maple syrup and then dredge them in the cornmeal mixture. Shake off any excess cornmeal right before frying.

3. Cover the bottom of a medium skillet with ¼ inch of oil. Heat the oil over medium-high heat until very hot. Fry the oysters until golden brown, 1 to 2 minutes per side. Drain on paper towels and serve with the sauce.

SERVES 2-4

Broiled Oysters

Don't tell guests how easy this delicious recipe is, and they'll never know.

4	tablespoons (½ stick) butter
½	teaspoon finely minced garlic
12	oysters, shucked and placed on the half shell
¾	cup coarsely grated Parmesan cheese
	Lemon wedges, for garnish
	Hot sauce

1. Preheat the broiler.

2. Melt the butter in a small saucepan over medium heat, then add the garlic, turn off the heat, and let the pan sit for 10 minutes.

3. Lay the oysters in a broiler pan, add ½ teaspoon garlic butter to each oyster, and top with 1 tablespoon of Parmesan. Broil for 3 to 5 minutes, until golden. Serve with lemon wedges and hot sauce.

SERVES 2-3

CONNECTICUT OYSTERING

Oysters thrive in Connecticut's tidal rivers and along Long Island Sound, and with 250 miles of shoreline, oyster farming became a major industry during colonial times. Farming for wild oysters was so popular that by 1750, the General Assembly allowed shore towns to regulate shellfishing. New Haven voters passed a law in 1762 prohibiting oystering from May through August, when they were spawning, to allow them time to mature and reproduce. By 1858, 250 schooners were importing 2 million bushels to New Haven's oyster shops, where they were then shipped inland. While Connecticut saw the same decline in the 20th century as other states did, today Connecticut aquaculture (farming aquatic plants and animals) is one of the United States' fastest-growing agricultural industries, with big beds in Mystic, New Haven, Bridgeport, and Norwalk. At regional farmers' markets, there is an abundance of vegetable and fruits in the warmer months from this fertile valley, and seafood from clams to quahogs year-round.

Clams *WITH* Chouriço *AND* Kale

This simple dish takes advantage of the clams you've just dug (or bought) and becomes supper with salad and grilled garlic crostini. Chouriço is a smoked Portuguese sausage that is mild and fat (like a thick hot dog) — not to be confused with the squat, spicy Mexican sausage called chorizo.

1	tablespoon grapeseed oil
½	small onion, finely diced
½	teaspoon minced garlic
¾	cup diced chouriço
4	cups coarsely chopped kale, stems removed, lightly packed
1	cup clam broth
20	clams
	Garlic Crostini, for serving (recipe follows)

1. Heat the oil in a large pot over medium heat. Add the onion, garlic, chouriço, and kale. Toss to coat with oil, and cook for 5 minutes.

2. Add the clam broth and cook for 20 to 25 minutes, or until the kale is tender. Add the clams, cover, and cook until the shells open, 5 to 8 minutes. Discard any clams that don't open. Spoon into bowls with broth, and serve with garlic crostini.

SERVES 2

Garlic Crostini

¼	cup extra-virgin olive oil
2	garlic cloves, minced
	Salt
½	baguette, sliced diagonally

1. Preheat the broiler.

2. Combine the olive oil, garlic, and salt to taste in a small bowl, and brush onto both sides of the bread slices.

3. Broil 1 to 2 minutes per side, or until toasted.

SERVES 2

Grilled Littlenecks with Garlic White Wine Butter

With a briny, buttery perfection, grilled littlenecks are an ideal outdoor summer appetizer. You can serve them with hot sauce (I've always said anything salt can do sauce can do better), though it's not necessary.

6	tablespoons unsalted butter
2	teaspoons minced garlic
2	tablespoons white wine
24	littleneck clams, well-scrubbed

1. Prepare a medium-hot fire in a gas or charcoal grill.

2. Melt the butter in a small saucepan; add the garlic and sauté until fragrant. Add the wine, stirring, and simmer for 2 minutes. Keep the sauce warm.

3. Place the clams directly on the grill, cover, and cook until the clams open, about 8 minutes. Spoon a teaspoon of butter sauce into each clamshell and grill an additional minute. Remove the clams with tongs, taking care not to spill the sauce in the shell, and serve immediately.

SERVES 4-6

CLAMMING

You don't need much more than a sense of adventure:

- A clam rake (I use a spoon or a garden hoe in a pinch)

- A basket or bucket (a beach towel will work if you're unprepared)

- A shellfish permit. Tidal flats are a public resource, and can be harvested by anyone with a permit. You can usually get a temporary one — or at least get the skinny on it — at the local town hall. If you're a vacationer or a non-resident, expect to pay twice what a local would pay for your license.

Once you've dug the clams, now what? First, discard any that are opened. Hard-shell clams just need to be rinsed off before cooking or serving, but soft-shell clams (steamers) should be rinsed several times under cold water, then immersed in a seawater or salted water (a tablespoon of salt for every quart) for at least an hour to remove any sediment before cooking.

(If you're not cooking for a few hours, it's convenient to leave them in a bucket of salt water in a cool place for up to eight hours. If the water gets murky, feel free to change it — the clams are simply discharging their sand and grit. Don't keep clams submerged in water for an extended period. If you aren't cooking until the next day, store in the refrigerator or on a bed of ice in a cooler.)

NEVER A BAD DAY

A bad day of clamming is still a great day. It's you, the wind, the water, maybe a friend, and intense digging. It's clams squirting back at you when you least expect it. Your bathing suit gets filthy (because you ultimately have to sit down) and your nails are ruined, but who cares. It's solitary. It's camaraderie; some of the best conversations you'll have with girlfriends, parents, boyfriends, and kids. There's something timeless about a vast stretch of untouched beach, the sun beating down on your back (or the rain coming). Silence. Memories of childhood. The wind. The tide. The harsh colors, the squinting in the sun. You can feel the history in the soil you dig, the remembrance of Native Americans and ancestors past. Once you're finished, and savor the clams that night with butter, you tell your tales and can't wait to get back out there.

Linguine *with* Clams

Like a favorite song, this is an old standard — an easy, go-to recipe for a cozy winter's night by the fire, remembering the sea. This recipe also provides a good opportunity to experiment with razor clams.

¼	cup olive oil
2	garlic cloves, minced
2	tablespoons white wine
1½	cups clam broth
½	cup chopped fresh parsley
2	tablespoons basil chiffonade (see page 53)
½	teaspoon salt
1	pound linguine
¾	cup minced clam meat
⅓	cup grated Parmigiana Reggiano plus more for garnish

1. Bring a large pot of salted water to a boil.

2. Heat the oil in a skillet over medium heat. Add the garlic and sauté until lightly browned, 1 to 2 minutes. Add the wine and cook for an additional minute. Add the broth, parsley, 1 tablespoon of the basil, and the salt, and cook for 5 minutes.

3. While the sauce is cooking, cook the linguine until just al dente. Drain and keep warm.

4. Add the clams and Parmigiana Reggiano to the sauce, stirring, and cook for 5 minutes longer. Serve over linguine, garnished with additional cheese and the remaining 1 tablespoon basil.

SERVES 4

Ipswich Clams *with* Garlic, Fennel, *and* Romesco Sauce

Shellfish can be steamed or poached with beer, wine, or water. Adding aromatics to the liquid — garlic, shallots, fresh parsley, rosemary, thyme, bay leaf — enhances the flavor. This dish is delicious with or without Romesco sauce.

1	tablespoon olive oil
1	leek, thinly sliced
½	fennel bulb, thinly sliced
2	garlic cloves, minced
3	Roma tomatoes, medium dice
1½	cups white wine
1	teaspoon salt
¼	teaspoon black pepper
¼	teaspoon red pepper flakes
¼	cup Romesco Sauce (recipe follows)
24	hard-shell (littleneck) clams

1. Heat the oil in a large skillet over medium heat. Add the leek and fennel and sauté until soft, about 15 minutes.

2. Add the garlic and sauté an additional minute. Add the tomatoes, wine, salt, pepper, and pepper flakes, and simmer, covered, for 30 minutes, or until the fennel is very tender.

3. Stir in the Romesco sauce and add the clams. Cover and cook until the clams open, 5 to 10 minutes. Discard any unopened clams.

SERVES 4 (APPETIZER) OR 2 (MAIN COURSE)

Romesco Sauce

When I had Romesco sauce for the first time, I was shocked — why had no one told me about this Spanish sauce? Like a pungent red-pepper pesto, Romesco is creamy, nutty, and versatile, a traditional sauce that's making a comeback in restaurants nationwide. Swirl it into seafood stews, spoon it over vegetables, or use it to dress simply prepared fish. Made with roasted red peppers, garlic, and nuts, it can be served over almost anything. In this version, there's a whisper of heat in your throat from the cayenne. If you like a bit more tingle, double the amount.

2	large jarred roasted red peppers (8 ounces)
1	garlic clove
¼	cup toasted slivered almonds or pine nuts
½	small yellow onion
½	cup roasted tomatoes (page 293)
2½	tablespoons sherry or red wine vinegar
2	teaspoons smoked paprika
3	tablespoons flat-leaf parsley
½	teaspoon cayenne, Aleppo pepper, or red pepper flakes
¼	cup plus 2 tablespoons olive oil
	Salt and freshly ground black pepper

1. Put in a food processor the peppers, garlic, almonds, onion, roasted tomatoes, sherry, paprika, parsley, and cayenne. Purée until blended and partially smooth.

2. With the motor running, slowly pour in the oil and continue blending until smooth. Season with salt and pepper to taste. Covered and refrigerated, Romesco sauce will last a week.

MAKES 2 CUPS

Rhode Island Clam Cakes

A masterful street food, Rhode Island clam fritters (aka clam cakes) aren't cakes at all — the best are briny Southern hush puppies. Think clam beignet; crispy and golden brown on the outside, pillowy and light inside, with savory bits of chewy minced clams and steam rising from the first bite.

I was introduced to them as a counter girl at the Kool Kone in Wareham on Route 6, a tired, narrow strip of road that once was the main drag from New York to Cape Cod. ("Is this the way to Wareham?" the old yaw goes. "I don't know, ma'am, but they look all right to me.") My first week on the job, making an ice cream sundae, I spilled a gallon bucket of walnuts on the filthy floor. In tears, I apologized to my bosses, an elderly couple who worked like dogs all summer (he was the jokester at the fryolator, she was the bad cop at the window tending the girls), then packed up their motor home and drove to Florida for the winter. "Don't worry," she said, chewing spearmint gum between her front teeth, "You'll rinse them, dry them off, and put them back in the bucket."

Serve the clam cakes with Tabasco sauce, lemon, and tartar or rémoulade sauce. They refry well the next day, too: just pop them in hot oil for a minute or two.

1	tablespoon fresh chives
1	tablespoon fresh chopped parsley
2	teaspoons Old Bay seasoning
1	teaspoon salt
¼	teaspoon black pepper
2	cups cake flour
1	teaspoon baking powder
1	teaspoon minced garlic
	Pinch of cayenne pepper
2	eggs
½	cup beer (needn't be fancy)
1½	cups minced clams
½	cup clam broth
	Vegetable oil for frying

1. Mix together the chives, parsley, Old Bay, salt, pepper, flour, baking powder, garlic, and cayenne. In a large bowl, whisk together the eggs and beer, then add the clams and broth. Mix in the dry ingredients and stir to combine.

2. In a heavy skillet, heat oil (2 to 3 inches deep) to 360°F (185°C). When the oil is hot, drop a tablespoon of batter at a time and fry until golden brown, 3 to 5 minutes per side. Fry just a few at a time. Remove with a slotted spoon and drain on paper towels.

SERVES 4-6

Stonington Clam Cakes *with* Lemon-Parsley Sauce

This is a forthright recipe for a summer appetizer or fall lunch. Serve on a bed of Boston lettuce.

LEMON-PARSLEY SAUCE

¼	cup mayonnaise
¼	cup sour cream
2	tablespoons lemon juice
2	tablespoons chopped parsley
1	teaspoon Dijon mustard
	Salt and freshly ground black pepper

CLAM CAKES

1	spicy Italian sausage link (casings removed), cooked and crumbled
1	cup chopped clams
1	egg, beaten
¼	cup all-purpose flour
2	tablespoons chopped parsley
½	teaspoon salt
½	cup vegetable oil or more
1	cup panko breadcrumbs (unseasoned)
	Pinch of cayenne
	Lemon wedges, for garnish

1. To make the sauce: Mix the mayonnaise, sour cream, lemon juice, parsley, mustard, and salt and pepper to taste in a small bowl. Adjust the seasonings.

2. To prepare the clam cakes: Combine the sausage, clams, egg, flour, parsley, and salt in a bowl. Mix well, and then separate the mixture into four patties.

3. In a skillet large enough to hold the clam cakes, heat the oil over medium-high heat. (You want the oil to be ¼ inch deep, so you may need to add more depending on the size of your skillet.) Season the breadcrumbs with cayenne and place on a plate. Coat the cakes on both sides with crumbs. Gently place the clam cakes into the hot oil and cook, turning once, until golden, 5 to 7 minutes per side. Remove with a spatula and drain on paper towels. Serve on a platter or individual plates with lemon wedges and the sauce.

SERVES 4 AS AN APPETIZER

Clams al Fresco with Tomatoes

This recipe is all about summer. Outdoors. On the deck. White wine. Beer. Pandora. Barefoot. The tomato sauce simmering gently with the juice of fresh clams, garlic, and olive oil. Eat it for a first course, or main course, or the next morning for breakfast. Served over pasta, just call it red clam sauce.

- 1 tablespoon olive oil
- 2 cloves garlic, minced
- 1 (28-ounce) can crushed tomatoes
- 1 tablespoon chopped basil
- ¼ teaspoon salt
- 8 ounces linguine or spaghetti
- 20 littleneck clams
- ¼ teaspoon red pepper flakes

1. Heat the oil in a skillet over medium-low heat. Add the garlic and cook for 30 seconds. Add the tomatoes, basil, and salt. Bring to a simmer and cook for 20 minutes.

2. While the sauce simmers, fill a pot with water, bring to a boil, and cook the pasta according to the directions on the package. Put a cup of water in a separate pot and bring to a boil. Add 10 of the clams, cover, reduce the heat, and cook until the shells steam open, 8 to 10 minutes. Discard any shells that don't open. Cool the clams, remove the meat from the shell, and chop.

3. When the sauce is cooked, add the remaining whole raw clams to the skillet and sauté until they open, 8 to 10 minutes. Add the chopped clams and the pepper flakes to the sauce, stir, and serve over the pasta.

SERVES 4

SUMMER MEMORIES

In college in the 1940s, my mother and her four sisters waitressed summers on Nantucket, and one evening she was sure she was waiting on Fred Astaire. She asked him, and the man demurred, saying he was often mistaken for the famous actor. When he left, she and her sister Patty were on a smoke break out in the alley, watching him depart. Down the street, he launched into a dance routine from "You'll Never Get Rich," then looked over his shoulder and smiled back at them before walking away.

OLD SALTS

Men who love the sea inevitably work their way up to a serious boat, and my father's first was a Herreshoff.

Though I was only four, I still remember it, a 21-foot Islander with a narrow beam and beautiful lines. He bought it with his friend Chuck Russell and they kept in the Weymouth Back River at the South Shore Yacht Club. Built in the U.K. in 1953, it was designed by Sidney Herreshoff, son of Captain Nathanael Herreshoff, from Bristol, Rhode Island, who started building boats in 1878 with his brother John. The Herreshoff brothers designed and built racing sloops that won five America's Cups,

and Nat Herreshoff ("the Wizard of Bristol") is considered one of the greatest boat designers of all time. (Incidentally: in 1851, Britain challenged the U.S. to a 53-mile yacht race. When America's schooner, called *America*, won by 8 minutes, Britain accused the U.S. of cheating. *America* was declared the winner, and the regatta was renamed America's Cup. The U.S. went on to win the race for 132 years, until Australia won in 1983. Every single winning America's Cup boat from 1893 to 1934 was built in the Herreshoff boatyard.)

Our boat was named *Queequeg*, after the cannibal harpooner full of derring-do in *Moby-Dick*, and my father won a slew of races in that boat just 20 miles from New Bedford, where Queequeg became fast friends with a wandering sailor called Ishmael.

Once, my mother had an argument on the boat with my father's friend Chuck (sort of a know-it-all physicist, but a smart guy whose creds were emboldened by the fact that he'd worked on the Manhattan Project), and she got so steamed up that she drew an imaginary line down the center of the boat, told him the port side was hers, the starboard was his, and not to cross the line. My father ended up laughing so hard that she cracked up laughing, too, and they all made up. They were like that.

Stuffies

Back when pickle makers weren't cool, and vintage meant grapes, you could get stuffies at The Blue Flame, a dive bar near the water in Onset. Stuffed quahogs, fondly known as "stuffies," are a specialty in southeastern Massachusetts and Rhode Island. Typically a mixture of breadcrumbs, Tabasco sauce, minced onions, celery, peppers, quahogs, and Portuguese sausage, there are nearly as many stuffing recipes as there are locals. When I'm at the beach, I make them at least once a week, quahogging at low tide with friends and then cooking them up before dinner.

8	quahogs (hard-shell clams)
4	tablespoons butter
2	garlic cloves, minced
½	medium yellow onion, minced
½	red bell pepper, diced
8	ounces minced chouriço sausage
	Juice of 1 lemon
	Salt and freshly ground black pepper
1	cup breadcrumbs
2	tablespoons minced fresh basil or parsley

1. Preheat the oven to 350°F (180°C).

2. Fill a big pot with 2 inches of water and bring to a boil. Add the clams, cover, reduce the heat, and cook until the shells open, 8 to 10 minutes. Discard shells that don't open. Remove clams from the pot with tongs, reserving the broth.

3. When the clams are cool enough to handle, remove the meat and break the hinge, saving 12 shells (you may not use them all). Put the shells on a baking sheet, open-side up. Finely chop the meat.

4. Melt the butter in a skillet over medium-low heat. Add the garlic and sauté until fragrant, about 1 minute. Add the onion and bell pepper and sauté until soft. Stir in the reserved clam meat. Add the sausage, stirring until warm. Add lemon juice. Season with salt and pepper to taste.

5. Remove from the heat and gently stir in the breadcrumbs and basil. Add enough clam broth to turn the mixture into a thick paste that holds together. Divide the mixture evenly among the shells, mounding them slightly. Bake until golden brown, 15 to 20 minutes. If you wish, you can toast them under the broiler for an additional minute.

SERVES 4-6

Seafood Fideua (aka Paella)

Paella ("pan" in Spanish) is a saffron-colored rice dish enhanced with seafood or meat. Since the 1920s, Spaniards in the Valencia region have substituted thin noodles for rice and called it fideua (FID-a-wah), from the Catalan word for noodle: *fideu*. While there are many variations, it is often made with white-fleshed fish and shellfish; in this version, shellfish gives the nutty-flavored pasta the briny flavor of the sea. Annatto seeds flavor and color the oil, lending the pasta its brick-colored hue.

¼	cup olive oil
1	tablespoon annatto seeds
4	garlic cloves, sliced
1	pound capellini or angel hair pasta
1	pound shrimp, peeled and deveined (see page 28)
1	pound scallops
1	pound clams, scrubbed clean
1	pound mussels, scrubbed clean
2–3	cups clam juice, warmed
2	tablespoons fresh parsley, basil, or cilantro, finely chopped

1. Heat the oil over medium heat in a skillet or pan large enough to accommodate all the seafood. Add the annatto seeds and sauté for 2 minutes, taking care not to burn the oil or seeds. Remove from the heat and let steep for 15 minutes.

2. Discard the seeds and reheat the oil over medium heat. Add the garlic and cook until soft and fragrant, about 1 minute.

3. Break the pasta in thirds and drop into the simmering oil. Stir constantly until lightly toasted. Add the shrimp and scallops to the pan and sauté until the shrimp just begins to turn pink, a few minutes. Add the clams and mussels, pushing them down to make contact with the hot pan.

4. Add enough clam juice to just cover the pasta and bring to a simmer. Continue to push the pasta down to keep it submerged under the broth. When the pasta is tender and the seafood almost cooked, lower the heat and continue cooking until all the liquid has evaporated and the pasta begins to crisp.

5. Discard any clams or mussels that have not opened. Sprinkle the parsley over the top and serve.

SERVES 6-8

SHIPWRECKED ON DUXBURY BEACH

On my first boat delivery from Southwest Harbor, Maine, to Virgin Gorda, we were cruising down the Atlantic coast on a beautiful 40-foot Hinckley sloop on an October night. We gave the city lights of Boston a wide berth on the leeward rail, sailing south toward Plymouth and the Cape Cod Canal, which would deposit us into Buzzards Bay, then Narragansett Bay by dawn. We'd spent the afternoon shucking clams on deck, and Liz, the cook, had made some clam chowder, which was simmering on the gimbaled stove.

The night was a little cloudy, but we could see a beacon of light that we thought were the lights marking the canal.

These were the days before GPS, when sailors steered by nautical charts and parallel rules. We realized our error too late; moments after we heard the crashing surf we slammed onto Duxbury Beach, where Myles Standish had settled over 300 years before. It was high tide, though by the time we organized ourselves and the Coast Guard was alerted, the tide had gone out. We spent an embarrassing night high and dry on Duxbury Beach, awaiting the morning tide when the Coast Guard, with the help of the tidal tow, could pull us off the beach. Our nerves shot, we cracked open a few beers and enjoyed the clams.

Nantucket Littlenecks

Thirty miles south of Cape Cod, Nantucket (Algonquin for "faraway land") is one of those places where you can squint your eyes and think you're living in the late 18th century. The island is so primly intact that it was designated a national historic landmark district by the National Park Service in 1966. Descendants of Tristram Coffin (one of the early colonial settlers), my mother's family still had reunions on the island with hundreds of Coffins (people were dying to get in when I was young).

1	cup dry vermouth or white wine
1	cup chopped yellow onion
2	tablespoons chopped parsley
36	littleneck clams, cleaned
4	tablespoons butter

1. Bring the vermouth, onion, and parsley to a boil in a large pot. Add the cleaned clams. Cover and steam until the shells open, 7 to 8 minutes. Begin checking after 5 minutes, transferring any opened clams to a big bowl. Continue checking, transferring the remaining clams (discarding unopened ones) with a slotted spoon to the bowl.

2. Strain the broth into a saucepan and bring to a boil. Reduce by half, then lower the heat and add the butter slowly, until melted. Divide the clams among six shallow bowls and pour the sauce over the clams.

SERVES 6 AS AN APPETIZER

WHAT MAKES A COASTAL TOWN COASTAL?

I think it's the boats. In the water. Out of the water. In gas station parking lots. In front yards. In side yards. In driveways. Towed down Main Street. Stopping traffic. Slowing traffic. Shrinkwrapped. Cradled. Gleaming. Painted. Rusty. Peeling. Soggy. Bobbing. Rocking. Standing at attention. As much a part of the landscape as the ebb and flow of the tides. It's definitely the boats.

Steamers *with* Lemon Butter

On Sunday nights in the summer, Uncle Jack would bring home a bushel of steamers, and we'd steam them up and sit around the round table and eat them — nothing else — for dinner, dipping the drippy morsels into hot broth then melted butter. Afterwards we'd sip the broth, which tasted of the sea. The brininess of the clams combined with hot butter is intoxicating. It's all you need for a real taste of New England.

5	pounds soft-shell clams
1½	sticks unsalted butter, melted
	Juice of 1 lemon (optional)

1. Fill a large, tall pot with an inch of water and place a steamer rack at the bottom (if you don't have one, don't worry about it). Place the steamers carefully on the rack (no more than 5 deep), and bring the water to a rolling boil. Cover and steam until the shells open, 7 to 8 minutes. Begin checking after 5 minutes, transferring any opened clams to a big bowl. Continue checking, transferring the remaining clams (discarding unopened ones) with a slotted spoon to the bowl, reserving the hot broth.

2. Heat the butter in a small saucepan over medium heat. Squeeze the lemon juice, if desired, into the hot butter, and pour into a few ramekins for shared dipping. Pour the reserved broth into mugs for dipping and drinking.

3. To eat the clams, pull the clam from the shell. Peel off the black membrane. Holding the clam by the foot (it looks like a neck), swish the clam in the broth to remove any sand or sediment, then dip into the melted butter and eat (neck, too, if you wish).

SERVES 6

VARIATION

Melt 2 tablespoons butter in the bottom of the pot and sauté 2 cloves minced garlic. Add the clams and enough white wine or vermouth to steam them in ½ inch of liquid, plus a tablespoon of fresh chopped parsley. Cover with lid, bring to a boil, and steam the clams as described in step 1. Serve the clams in their juices with hot, crusty bread.

Steamed Clams in White Wine, Butter, and Garlic

Sometimes you just want honest food. No fuss. Steamed littlenecks with garlic, butter, and wine are about as delicious as it gets, and as simple as it sounds. Serve with wine and hot, crusty bread for sopping.

- 2 tablespoons butter
- 2 tablespoons extra-virgin olive oil
- 3 garlic cloves, minced
- ½ cup dry white wine
- 32 littleneck clams, rinsed and scrubbed
- 2 tablespoons fresh chopped parsley
- 1 lemon, quartered

1. In a stockpot with a tight-fitting lid, melt the butter and oil over medium-low heat, add the garlic, and cook until the garlic is fragrant, about 2 minutes.

2. Raise the heat to medium-high, add the wine, stir, and bring to a boil. Reduce the heat to a gentle simmer. Gently add the clams, cover, and steam until the clams open, 3 to 5 minutes. Discard any clams that don't open.

3. Transfer the clams and steaming broth to serving bowls, sprinkle with parsley, and serve with lemon and cute little forks if you have them.

SERVES 4 AS AN APPETIZER

"The Shoreman does not claim to be a philosopher . . . .
He believes, however, that affairs would be settled more
quickly if all men had an opportunity to dig a few clams."

— HAYDN SANBORN PEARSON, "SEA FLAVOR," 1948

LITTLENECKS

CHERRYSTONES

STEAMERS

QUAHOG

TYPES OF CLAMS

It can be confusing. Clam is a term given to a number of different species of shellfish called bivalves.

Atlantic hard-shell clams are distinguished by size. The largest hard-shell clams are called quahogs and were used to make wampum by Native Americans. They are usually too tough to eat raw, but are great in chowders and stuffed. Cherrystone clams (2 to 3 inches in diameter) are tasty steamed or grilled, or delicious served raw on the half shell. Littlenecks (1 to 2 inches) are tender and juicy, excellent raw, steamed, or cooked. (Mahogany clams are small, hard-shell quahogs similar in size to littlenecks. They are harvested in the deep waters of Maine.)

Hard-shell clams differ from soft-shell clams — aka steamers, Ipswich clams, piss clams (you'll see why), long-neck clams, or fried clams — which are always served cooked. Their shells are brittle (if you squeeze hard they'll break), and their neck sticks out when they are out of the water. You'll find them in saltwater sandbars and mudflats. They live in the sand just below the surface, and at low tide they squirt water and leave a little hole behind in the wet sand when you walk by. When you see them squirt (sometimes you can pound along the wet sand and cause them to squirt), you have to dig like mad to get them.

A razor clam is a sharp-edged "soft-shell" clam whose shape bears resemblance to a straight-edge razor. Like a steamer, the shell won't close completely when out of water. Eaten cooked, razor clams are considered a delicacy, which cracks me up, since we used to chase them into their holes on the beach at low tide.

Clams should always be alive and closed up tight when purchased and cooked. If the shell is open, tap it — if the clam's alive, it will clam up. If not, discard it. Store on ice or in the refrigerator. Don't put clams in a plastic bag, since they need to breathe.

New Haven–Style
White Clam Pizza

On a small street not far from New Haven's waterfront, a simple Italian pizzeria has been turning out "tomato pies" for almost a hundred years. In the 1920s, Frank Pepe used a coal fire to give his pies their famous crisp, chewy crust. His nephew would shuck the fresh clams he'd collected on the shore and sell them on the half shell. Someone got the brilliant idea to put them on the pizza, and this pizza was born. Today Frank Pepe's sells only pizza, and one of their specialties is White Clam Pizza. They get fresh whole clams delivered daily (if the shipment doesn't come in, they don't serve the pizza) and one guy's job is just to shuck clams. This adaption is based on Frank's recipe.

DOUGH

¾	cup warm water (110–115°F [43–46°C])
1½	teaspoons active dry yeast
2¼	cups all-purpose flour plus more for kneading
1	teaspoon salt
2	tablespoons olive oil plus more for oiling the bowl
1	tablespoon cornmeal, for dusting the paddle

TOPPING

¾	cup chopped fresh clam meat plus 1 tablespoon liquid
2	tablespoons extra-virgin olive oil
½	teaspoon minced garlic
1	teaspoon fresh oregano
¼	cup grated Parmesan cheese
¼	cup grated Romano cheese
1	cup fresh baby arugula leaves

1. To make the dough: Put the warm water and yeast in a small bowl, whisk, and set aside for 10 minutes to proof. Combine the flour and salt in a large bowl; make a well in the center, and pour the yeast mixture and oil into the well. Using your fingers or a fork, gently incorporate the flour into the liquid. Turn the dough onto a lightly floured surface and knead until smooth, about 5 minutes. Place into a lightly oiled bowl, cover with plastic wrap, and set aside in a warm, dry area until the dough doubles in size, about 2 hours.

2. Preheat the oven to 500°F (250°C). Put a pizza stone on the lowest level.

3. To make the topping: Drain the clams, retaining 1 tablespoon of liquid. Mix the chopped clams and liquid with the oil, garlic, and oregano in a small bowl.

4. Turn the dough onto a lightly floured surface and roll into desired shape. Dust a pizza paddle with cornmeal and place the dough on it. Spread the clam mixture evenly over the surface, and then sprinkle with the Parmesan and Romano cheese. Slide the pizza onto the hot stone and cook until golden brown, about 10 minutes. Cut the pie, topping each slice with a sprinkling of baby arugula, and serve.

MAKES 1 PIZZA (SERVES 4)

WINKLES

Pull up a patch of wet seaweed near the water line on a rocky tidal beach and you'll probably find periwinkles, those cute spiraling shore snails that close up their hatch and retreat into their shell when you lift them off a rock they've been clinging to. Tiny, tasty periwinkles. Simply put them in a pot of boiling water with salt and cook for 7 minutes. Drain them, remove the meat with a lobster pick or toothpick, and eat them. Some people bring the mouth of the shell to their lips and suck; you'll get a rush of briny juices, then you can scoop out the flesh with a lobster pick. What fun to sit around, drinking beer, sucking winkles. You can also broil them with garlic butter like their fancy French cousin, escargot.

Largely overlooked — somewhere below pigeon on the popularity scale — winkles are sweet and easy to find, plus you don't need a permit to collect them. Native to Europe, periwinkles were brought to Halifax, Nova Scotia, in the 1850s as ballast aboard ships, and from there migrated south along the Atlantic seaboard.

Whelks *with* Parsley *and* Garlic Butter

When we collected whelks in our buckets as kids, it never occurred to us that they were *edible*. Mostly I enjoyed playing with their dried egg casings — the spiraling rattling hoses that we called "mermaid's necklaces." Tasting like a cross between a steamer and a mussel, whelks are a sea snail (a gastropod) that's netted on Nantucket Sound, Vineyard Sound, and Buzzards Bay. They're often mislabeled conchs, which are close cousins. Whelks are carnivores that grow in temperate zones, whereas conchs (those white spiral shells you find on Caribbean beaches) are herbivores that grow in tropical waters.

In New England, more than a few fishermen turned to conching, as some of them call it, as cod and haddock became difficult to fish. On Martha's Vineyard, where fishermen landed over a million pounds of whelks in 2012, whelks have become a lucrative export.

1	pound whelks (10–12), scrubbed
½	cup (1 stick) butter
2	garlic cloves, minced
¼	bunch Italian parsley, chopped
	Salt and freshly ground black pepper
	Cayenne pepper
	Lemon

1. Soak the whelks in cold water for an hour to clean them.

2. Bring a pot of heavily salted water to a boil, add the whelks, reduce to a simmer, and simmer for 10 minutes. When slightly cooled, remove each whelk from its shell (pinch off the disc that looks like a trap door), and take off the dark gray sack at the bottom of the whelk. Rinse and dry the shells and set them aside.

3. Preheat the oven to 350°F (180°C). Melt the butter with garlic in a pan over low heat, then add the parsley, salt, pepper, and cayenne. Turn off the heat and add the whelk meat, stirring to coat; then push the meat back into the shells.

4. Lay the shells on their backs in an ovenproof dish, drizzling the leftover butter on them, and bake for 5 to 10 minutes. If they won't stay put, lay them on a bed of salt. Serve with lemon squeezed over them.

SERVES 4 AS AN APPETIZER

ONE SUMMER IN MAINE

Several years ago we spent a few weeks on Mount Desert Island, staying in Pretty Marsh. One day, we canoed around Bartlett Island, owned by the Rockefellers, and came across a tiny island — a speck on the chart — whose small circular shore was literally covered with mussels. There were so many mussels I was sure cockles weren't far behind, nor Molly Malone.

My kids thought I was nuts, but we returned to the island every day, buckets in hand, and picked enough to feast on. Everyone else got pretty sick of mussels, but to me it was a bonanza. That's the thing about seafood: it's best when it's fresh, and it won't get any better than it is right now. It begs to be eaten.

Steamed Mussels *with* White Wine Sauce *and* Garlic-Saffron Toasts

Like other bivalves, mussels are filter feeders that cleanse the water they reside in, stocking up on valuable omega-3 fatty acids from algae along the way. Steamed mussels can be enjoyed any number of ways — cooked with wine and celery, served in a broth of garlic and tomatoes as they do in Spain, or steamed in beer as you might find in Charlevoix, along the St. Lawrence River. Years ago I visited a café in Charlevoix that served only mussels, *pommes frites*, and beer, with multiple options in each category; I was in heaven. This is an earthy, rustic dish; you really don't need utensils. Garlic toast sops up the juices.

1	tablespoon olive oil
2	shallots, minced
1	celery stalk, chopped fine
1	teaspoon minced garlic
2	cups dry white wine
½	teaspoon fresh thyme
½	teaspoon salt
	Freshly ground black pepper
2	pounds mussels, scrubbed and debearded
4	tablespoons butter
	Garlic-Saffron Toasts (recipe follows)

1. Heat the olive oil in a large pot over medium heat. Add the shallots, celery, and garlic, and sauté until fragrant, a few minutes. Add the wine, thyme, salt, and pepper to taste, and simmer 5 minutes.

2. Increase the heat to high, add the mussels, cover, and steam until the mussels open, 3 to 7 minutes. Check after 5 minutes and begin removing mussels as they open. Discard any mussels that never open. Divide the mussels among the bowls.

3. Add the butter to the pot and melt to make a buttery broth. Pour the broth over the mussels and serve with garlic crostini.

SERVES 4 (APPETIZER) OR 2 (MAIN COURSE)

Garlic-Saffron Toasts

1	small loaf French bread
½–¾	cup olive oil
2	cloves garlic, smashed lightly
1	teaspoon saffron threads
1	teaspoon lemon juice
½	cup store-bought or homemade mayonnaise (page 235)

1. Holding the bread on its side, cut diagonally into eight slices. Pour oil into a medium skillet to the depth of ⅛ inch, and heat on low. Add the garlic and sauté over low heat for 5 minutes, or until golden.

2. Remove the garlic, and brush the warm, garlicky oil on both sides of the bread slices. Over medium-low heat, toast the bread in the pan with the olive oil until golden brown on both sides, 5 to 7 minutes on each side. Transfer the slices to paper towels to drain.

3. In a small bowl, combine the saffron with the lemon juice, stir, and let stand for 5 minutes. Add the mayonnaise, stir well, and spread lightly on one side of each garlic toast.

SERVES 4

COOKING WITH MUSSELS

- Buy mussels that are deep blue-black and smell fresh, with closed shells. Tap the shells of any that are open. If the shell doesn't close, the mussel is dead and should be discarded (also toss any with broken shells). You want to cook your mussels when they are alive.

- If gathering mussels, check with the town hall about which beds are open, as a red tide can make them temporarily poisonous.

- Cook mussels the same day you catch or buy them.

- If they are really heavy, they may have sand in them — you can clean them in the same way you do soft-shell clams (page 41).

- To clean mussels you've caught, pull off the beards (the weedy growth) with your fingers — if stubborn, snip with kitchen scissors. Rinse well under cold water. (Farmed mussels don't have beards.)

ABOUT MUSSELS

We've come a long way since fishermen used mussels for live bait. Cooked extensively in Europe (Julia Child became a big fan in France), mussels have now become fashionable in the U.S., for good reason: easy to gather, these onyx gems are sweet and rich-tasting, high in protein, and a good source of vitamin C, folate, potassium, zinc, iron, phosphorus, and manganese. And did I mention vitamin B_{12}?

In New England, there are two types of mussels: blue and ribbed. With its blue-black shell and violet interior, the delicious northern blue mussel is found in eastern American waters north of the 35th parallel. Grown in Atlantic waters, they're what you'll find at a seaside fish market, and if you're lucky, in the wild, growing in clusters.

But they may not be there the next time you visit — mussel beds are targets for starfish and eider duck predators, and can be wiped out by storms. There used to be a huge natural mussel bed off Nantucket — until one winter a series of storms hit, winds blowing hard from the east all winter long, and one last storm smothered the bed.

Ribbed mussels are similar to blue mussels, but the shell is corrugated and more triangular. You'll find them clinging in shallow waters to clumps of mud at low tide in the mudflats. Ribbed mussels are critical to the health of a salt marsh, and often are rich in organic bacteria. They're also not as plump as blue mussels (which have been in deeper waters eating their hearts out). While not poisonous, they are not as tasty, and not often eaten by people.

According to the Cape Cod Commercial Fishermen's Alliance (their logo is "Small Boats — Big Ideas"), fishing for wild mussels is one of the last wild shellfish operations on the Cape and the Vineyard, with guys going out in small skiffs year-round, landing almost 800 pounds last year. Many mussels today sold in the U.S. are cultivated, with farmed fishing considered largely sustainable and environmentally friendly since shellfish filter water. From Sakonnet Point in Rhode Island to Penobscot Bay in Maine, some mussel farmers have also gone into bottom culture, leasing ocean floor and dropping mesh bags containing mussel seeds that hang nearly 30 feet below the water's surface. As the mussels grow, the bags deteriorate and they attach to a line.

Today, the United States imports 90 percent of the mussels sold. In Prince Edward Island, with its rural and fishing communities, mussel farming is second only to tourism (aquaculture farms on PEI exported $20 million of blue mussels to the U.S. last year), and many small New England fishermen and organizations are pointing to them as an example of what is possible in developing more offshore farms.

Chatham Mussels in Curried Ginger Coconut Broth

At Cape Cod's elbow, where tidal flats blur the distinction between land and sea, men can fix up an old skiff, get a license, and make a living shellfishing. For decades, the Chatham Shellfish Company bought steamers, conchs, and mussels right from the fishermen. Some of the company's old shanties at the bend of the Oyster River are still there, the wooden planks crusty with the white chalk from scores of oyster shells.

1	(13-ounce) can coconut milk
2	tablespoons grated ginger
1	garlic clove, minced
2	teaspoons curry powder
3	(1-inch) pieces lemon peel
	Juice of ½ lemon
¾	teaspoon salt
¼	teaspoon freshly ground black pepper
2	pounds mussels
2	tablespoons chopped cilantro

1. Combine the coconut milk, half a can of cold water, the ginger, garlic, curry, lemon peel and lemon juice, salt, and pepper in a large saucepan and simmer over medium heat for 20 minutes.

2. Add the mussels to the coconut broth, cover, and cook until they open, 3 to 7 minutes. Check after 5 minutes and begin removing mussels as they open. Discard any mussels that never open.

3. To serve, distribute the mussels among bowls, along with the broth, and garnish with cilantro.

SERVES 4 (APPETIZER) OR 2 (MAIN COURSE)

Saffron *and* Mussel Risotto *with* Shiitakes

In this Italian classic, you can substitute clams for the mussels, or other types of mushrooms for the shiitakes.

6	cups chicken broth
1	pinch saffron threads
3	tablespoons extra-virgin olive oil
3	garlic cloves, minced
½	yellow onion, minced
2	cups arborio rice
¼	cup dry white wine
¼	teaspoon salt
¼	teaspoon black pepper
2	tablespoons butter
8	ounces fresh shiitake or portabello mushrooms, stems removed and thinly sliced
20	mussels
2	tablespoons basil oil or olive oil
⅔	cup Parmigiana Reggiano, freshly grated in large chunks
2	tablespoons chopped flat-leaf parsley

1. Pour the broth into a saucepan, add a pinch of saffron threads, and simmer over low heat while you cook the risotto.

2. Heat 2 tablespoons of the olive oil in a large pot over medium heat and add the garlic, sautéing until fragrant, about 1 minute. Add the onion and cook, stirring, until soft and golden, about 10 minutes. Stir in the rice and coat all the grains with oil until the edges become translucent and there is a white dot in the center, 1 to 2 minutes. Pour in the wine, stirring the rice until the liquid evaporates.

3. Ladle 1 cup of broth into the pot, stirring, keeping the heat on medium-low. Season with the salt and pepper. While the risotto is cooking, heat the remaining 1 tablespoon olive oil and 1 tablespoon of the butter in a medium skillet over medium-high heat. Add the mushrooms, seasoning with salt and pepper, and sauté until browned, 10 to 15 minutes.

4. Back to the risotto: add another cup of broth, stirring. Continue to add the broth 1 cup at a time, stirring regularly as the water is absorbed. When three-quarters of the stock has been absorbed (roughly 14 minutes), add the mussels and the

rest of the broth and stir continuously until the mussels have opened (discard unopened ones) and all the broth has been absorbed by the rice. (If you have to add a little water, do so after testing the rice to see if it is done — you want it oozing and delicate, not too mushy, not too hard, and not too liquidy.)

5. Remove the risotto pot from the heat, stir in 1 tablespoon of the basil oil, the remaining 1 tablespoon of butter, and ⅓ cup of the cheese. Taste and season with salt and pepper. Ladle the risotto into shallow bowls and top with the mushrooms, parsley, and the remaining 1 tablespoon basil oil. Pass the remaining ⅓ cup cheese at the table.

SERVES 4

Ginger-Lime Scallop Ceviche

Ceviche is a Peruvian style of "cooking" seafood in citrus juice that probably started on the beach, when fishermen cut up their fresh catch and squirted lime juice over it. You needn't confine yourself to sea scallops — bay scallops work well, too, as do tuna and salmon. Just be sure to use extremely fresh scallops, and marinate them at least 30 minutes (for a "rarer" ceviche) and up to 3 hours (for one that is "fully cooked"), keeping the ceviche in the refrigerator until ready to serve.

You can present it in a bowl with tortilla or plaintain chips for a casual party, on a small bed of watercress or baby arugula for an appetizer, or in a stemmed martini or champagne glass for a special occasion such as New Year's Eve.

1	tablespoon chopped scallion greens
1	garlic clove, minced
1	tablespoon chopped cilantro
½	yellow bell pepper, seeded, small dice
½	red bell pepper, seeded, small dice
2	teaspoons minced fresh jalapeño
¼	teaspoon sea salt
1	(½-inch) piece fresh ginger, peeled and chopped
	Juice of 2 limes
1	tablespoon grapeseed oil
1	pound sea scallops, diced (if substituting bay scallops, just slice them in half)

1. In a bowl, combine the scallions, garlic, cilantro, bell peppers, jalapeño, salt, ginger, lime juice, and oil. Add the scallops to the bowl and toss gently, coating with the marinade. Cover with plastic wrap and refrigerate, stirring occasionally, for 30 minutes to 3 hours.

2. Sprinkle with salt and pepper to taste and serve.

SERVES 8

FISHY TO THE BACKBONE

The captain of a whaling ship had to be in charge. Nantucketers called a good captain — someone who was self-confident, didn't waver on decisions, was decisive, and loved killing whales — a "fishy man." "Fishy to the backbone" was the ultimate compliment a Nantucketer could receive. Sounds like a great recipe title.

Bay Scallop Ceviche

The size of large pearls, bay scallops are perfect for this centuries-old Peruvian method of "cooking" fish by marinating it in citrus juice.

6	ounces bay scallops, rinsed well in cold water and drained
1½	tablespoons diced scallion
1½	tablespoons diced red bell pepper
2	tablespoons chopped fresh cilantro
½	teaspoon grated orange peel
1	teaspoon finely chopped fresh habanero or jalapeño, seeded
2	tablespoons lime juice
2	tablespoons orange juice
	Sea salt and freshly ground black pepper
	Plantain chips, for serving

1. Combine the scallops, scallion, bell pepper, cilantro, orange peel, habanero, lime juice, and orange juice in a mixing bowl. Season with salt and pepper. Cover with plastic wrap and refrigerate, stirring occasionally, for 1½ hours. The scallops should be opaque on the outside, pink on the inside. (If you want them opaque inside, marinate for 3 hours.)

2. Spoon the ceviche into small glasses and serve with crisp plantain chips.

SERVES 2

BAY SCALLOPS

While sea scallops are the size of a half dollar, bay scallops are the size of a dime — a magnificent pearl stud earring. Fresh from local waters and cooked simply to accentuate the subtle, sweet, nutty flavor, bay scallops are sublime.

The bay scallop season is in full swing from November to April, although with eelgrass beds sensitive to pollution, the scallops aren't as prevalent anymore. But you'll find wild scallop beds on Buzzards Bay, the Vineyard, Nantucket, and the outer Cape. November 1 is opening day on the outer Cape; in Chatham, the inshore shallows are crowded with small boats hauling drags, trying not to bump into each other's lines and the winter sticks (moorings).

Unlike clams or mussels, scallops can't hold their shells closed, so it's important to shuck them quickly, before they lose moisture and die.

GINGER-LIME SCALLOP
CEVICHE

BAY SCALLOP CEVICHE

Herbed Baked Scallops

If you have any leftover baked scallops, lightly bind them with a white sauce, and you have a delicious crêpe filling.

- 2 tablespoons butter
- 1 pound bay scallops
  Salt and freshly ground black pepper
- 2 cloves garlic, minced
- 1 cup fresh breadcrumbs (3 slices bread, well toasted and chopped)
- 1½ tablespoons chopped parsley
- 1½ tablespoons chopped chives
- ½ tablespoon chopped fresh tarragon
  Smoked paprika

1. Preheat the oven to 450°F (230°C).

2. Brush an 8-inch square glass baking dish with a bit of the butter, and spoon the scallops into the dish in a single layer. Season well with salt and pepper.

3. Melt the remaining butter in a small skillet over medium heat. Add the garlic and sauté for 1 minute. Add the breadcrumbs and stir until they start to brown, 4 to 6 minutes.

4. Toss in the parsley, chives, and tarragon, stir to combine, and spoon the crumbs on top of the scallops. Dust smoked paprika over the top and bake until the scallops are opaque, about 13 minutes.

SERVES 4

FELICITY

My parents gave up sailing in their mid-seventies, primarily because my mother was worried about one of them falling overboard. After my mother passed away, my father said what the heck, and at age 78 went out and bought a honey of a boat, a Herreshoff 12½, also known as a Buzzards Bay Boy's Boat. It was his last of 32 boats over a lifetime, and what a boat; designed by the Wizard of Bristol (see page 166) in 1912 for the afternoon chop of Buzzards Bay, it's arguably one of the prettiest small sailing boats of all time. He'd sail her in Mattapoisett Harbor and out on Buzzards Bay, deftly bringing *Felicity* up to the mooring under sail with no motor, just like he'd done with boats for almost 80 years.

Pan-Fried Bay Scallops

The year we moved back East from Indiana, my father embraced the water with a vengeance. We sailed, swam, fished, clammed, canoed, and prepared for possible September hurricanes with vigor. As cold weather set in, we closed up the cottage, but still went scalloping on Buzzards Bay, my father having devised scallop goggles — essentially wooden boxes with glass at the bottom — that we'd float in the water near the eelgrass on the shore to find bay scallops in our waders. Then we'd rake like mad, and take them home to pan-fry in a cast-iron skillet with butter. Plain-frying allows the taste of the sea to shine through. Or you can add a few sprigs of thyme, or some garlic, or some white wine to deglaze the pan.

2   tablespoons butter
1   pound fresh bay scallops
     Salt and freshly ground black pepper

1. Melt the butter in a cast-iron skillet over medium heat. Toss in the scallops and cook for a few minutes, shaking the pan to flip them.

2. Turn off the heat and allow them to cook for a minute or two more.

SERVES 2

SOUTHEASTERN CONNECTICUT

The stretch of shoreline from the Connecticut River to Pawcatuck has a gravitational pull that far outweighs its population. Its seven towns — Old Saybrook, Old Lyme, East Lyme, Waterford, New London, Groton, and Stonington — account for a mere fraction of the state's 3.5 million residents, but Nutmeggers are unable to resist the beauty of this region that offers attractions ranging from Mystic Seaport to Stonington Vineyards. Interestingly, the lower Connecticut River discharging into Long Island Sound between Old Lyme and Saybrook creates big shifting sandbars at the mouth, leading the Connecticut River to become the only principal American river with no port at its mouth.

Pan-Seared Stonington Scallops
WITH Sorrel Sauce

A vegetable that sometimes disguises itself as an herb, sorrel has a distinctive tangy flavor, almost like lemon. Its closest cousin is spinach (or maybe mustard greens, if served raw), but that would be a different taste. When cooked, it turns an olive-y green, so reserve a few of the raw strips for a bright green garnish.

1	pound sea scallops
	Sea salt and freshly ground black pepper
2	tablespoons unsalted butter
2	tablespoons extra-virgin olive oil
1	garlic clove, minced
1	tablespoon white wine
	Juice of ½ lemon
2	tablespoons heavy cream
1	cup loosely packed sorrel plus ¼ cup for garnish

1. Rinse the scallops and pat them dry, then sprinkle with sea salt and pepper.

2. Melt the butter with the oil in a large skillet over medium-high heat. Add the scallops (flat) and cook until golden brown, about 2 minutes, then flip and sear the other side, making sure the scallops are cooked through, about 1 minute. Transfer the scallops to a plate and keep warm.

3. Add the garlic, wine, lemon juice, and cream to the skillet, stirring and scraping up any bits from the pan. Gather the sorrel bunch like a bouquet and cut across the leaves to make thin strips. Add 1 cup of the sorrel strips to the sauce (saving ¼ cup for garnish), stirring, and reduce for 1 to 2 minutes.

4. Divide the scallops among the plates and top with the sauce. Garnish with the reserved sorrel strips.

SERVES 4 (APPETIZER) OR 2 (MAIN COURSE)

Seared Ginger Sea Scallops

You don't need a professional kitchen with a Jenn-Air range to turn out good food. My cottage kitchen has an electric fourtop, a Formica counter, and Revereware I inherited from my mother. If I go all out, I might buy a nonstick skillet from Target when I need a new pan. When it comes to seafood, the home cook's inner self often fears failure and humiliation. I'm here to tell you: it's easy. And if the fish is fresh, it tastes best when it's simple, for it is then that you taste the flavor of the sea.

½	cup olive oil
2	tablespoons minced ginger
2	tablespoons lemon juice
2	tablespoons rice wine vinegar
¼	teaspoon salt
	Freshly ground black pepper
16	large sea scallops
4	lemon wedges

1. Combine the oil, ginger, lemon juice, vinegar, salt, and pepper to taste in a medium bowl. Add the scallops, toss to coat, and marinate for 30 minutes to 1 hour.

2. Set a dry cast-iron skillet over medium-high heat and heat for a few minutes. Remove the scallops from the marinade and sear 3 to 4 minutes per side. Serve with lemon wedges and encourage your guests to squeeze the juice over the scallops — it's delicious!

SERVES 4

Fried Scallops

When my cousin Doug came home to the Cape from Florida, he always headed first to a fast food joint and ordered a big plate of fried clams and scallops. To him, nothing tasted like "arriving" more than fresh fried seafood. I know what he meant. Sometimes it's what you want, dressed with nothing fancier than ketchup, tartar sauce, and a good coleslaw. This is for you Cod kids out there.

½   **cup white flour**
½   **cup cornmeal**
½   **teaspoon salt**
½   **teaspoon smoked paprika**
     Freshly ground black pepper
1½   **pounds sea scallops, medium sized**
     Vegetable oil

1. Combine the flour, cornmeal, salt, paprika, and pepper in a baking pan. Dry the scallops with a paper towel, then dredge all sides in the flour mixture. Shake off the excess flour.

2. Add enough oil to cover the bottom of a large skillet to ¼-inch depth. Heat the oil to 350°F (180°C) over medium-high heat, and then fry the scallops in batches, making sure not to crowd them, until golden brown, about 2 minutes per side. Drain on paper towels and serve immediately.

SERVES 4

NANTUCKET BAY SCALLOPS

The cold waters of Nantucket Sound and the protective inshore eelgrass found on the island of Nantucket 16 miles offshore make Nantucket Bay scallops particularly prized. Often diver-caught or raked, brought in live, they are shucked at dockside shucking shanties and known to be especially sweet and tender. While Nantucketers are allowed to harvest scallops in October, the commercial season begins November 1 and goes until March 1, but if the harbor freezes up you may not see them past January. If you can't find Nantucket Bay scallops, other bay scallops are a good substitute.

Seafood Quinoa Salad

Quinoa can be traced back over 6,000 years to the Aztecs in ancient Peru and to the Native Americans of North America. This salad features a traditional trio of seafood, but can be easily changed to suit any taste using different sea creatures. Refrigerate all seafood until ready to use.

SEAFOOD STOCK

	Reserved shrimp shells
5	cups cold water
1	bay leaf
1	teaspoon seafood seasoning, such as Old Bay

QUINOA

½	cup red quinoa
½	cup white quinoa
1½	cups seafood stock

VINAIGRETTE

2	tablespoons freshly squeezed lemon or lime juice
1	tablespoon finely minced shallot
	Salt and freshly ground black pepper
6	tablespoons extra-virgin olive oil

SEAFOOD

1	tablespoon olive oil
1	tablespoon unsalted butter
12	jumbo shrimp or ½ pound (31–35) smaller shrimp, peeled with tails left on for presentation (save the shells for the stock)
6	ounces white squid (calamari), cleaned and bodies cut in rings, tentacles included
6	ounces bay scallops (whole) or sea scallops (cut into pieces)
	Salt and freshly ground black pepper
1	teaspoon minced garlic
1	medium tomato, seeded and finely diced
1	tablespoon minced flat-leaf parsley
	Sea salt (optional)
1	lemon, cut in wedges, for serving

1. To make the stock: Put the reserved shrimp shells in a medium stockpot with the cold water, bay leaf, and seafood seasoning. Bring to a simmer and cook gently for 30 minutes, uncovered, until the liquid is reduced by half. Strain the broth into a large measuring cup, pushing down on the solids in the strainer to enhance the flavor.

2. To prepare the quinoa: Rinse the quinoa several times to remove the coating, which has a bitter flavor. Put the stock and red quinoa in a medium pot with a tight-fitting lid, and bring to a boil over medium heat. Simmer for 5 minutes and then add the white quinoa. Cover and simmer for 20 minutes, or until the quinoa starts to split open and the stock is absorbed. Taste for doneness. Spread out the grains on a rimmed baking sheet to cool.

3. To make the vinaigrette: Put the lemon juice and shallots in a large bowl. Season to taste with salt and pepper, and whisk in the olive oil in a steady stream to emulsify.

4. In a large skillet, heat the oil and butter for the seafood over medium heat. Season the shrimp, squid, and scallops with salt and pepper to taste. Add the shrimp, squid, and garlic to the skillet, cooking and stirring until the seafood is just opaque, about 3 minutes. Remove the mixture with a slotted spoon and place in the bowl with the vinaigrette.

5. Add the scallops to the skillet and cook until almost done, no more than 1 to 2 minutes. Seafood has a small window between cooked and overcooked, so err on the side of less. Add the scallops to the vinaigrette bowl.

6. In the skillet with the seafood fond (brown bits stuck to the bottom), add the quinoa to warm it, stirring to release the flavor. Place the quinoa in the bowl with everything else. Add the tomato and parsley to the seafood salad and mix gently. Sprinkle lightly with sea salt, if desired. Serve with lemon wedges.

SERVES 4 AS AN APPETIZER

New Bedford Scallops WITH Cashews AND Tomatoes

Remembering when New Bedford Harbor was polluted with PCBs, I was startled the first time I saw New Bedford scallops glorified on a menu. I soon learned: unlike Cuttyhunk oysters, say, which were probably caught or farmed just offshore this outermost Elizabeth island, "New Bedford scallops" means that a fishing trawler from this commercial port brought them in. They could have been dredged southeast of Nantucket or off Georges Bank.

This is an easy winter dish — tomatoes, nuts, and rosemary are a great flavor trio that you wouldn't necessarily connect, but they come together gloriously, dancing around the subtle mild bivalve at the center. This is adapted from a recipe in *Bon Appétit*, one of my favorite places to find quick, inventive dishes.

⅓	cup unsalted cashews, peanuts, pistachios, or a combination thereof
3–4	tablespoons olive oil
	Sea salt and freshly ground pepper
16	large sea scallops
½	shallot, finely chopped
1	clove garlic, minced
2	cups cherry tomatoes
2	teaspoons white wine or champagne vinegar
1	tablespoon chopped fresh rosemary

1. Toast the nuts in a toaster oven until browned and fragrant. Chop and toss with 1 tablespoon of the oil and sea salt and pepper to taste.

2. Pat dry the scallops and season with sea salt and pepper on both sides. In a cast-iron skillet, heat 2 tablespoons of the oil over medium-high to high heat almost to smoking and then add the scallops (you'll hear a sizzle when you add them). Cook the scallops until seared and cooked just through, about 3 minutes per side. Transfer them to a serving plate and cover with foil.

3. In the same skillet, use a wooden spoon to sauté the shallot, garlic (adding 1 additional tablespoon of olive oil if needed), and tomatoes, seasoning with salt and pepper, and scraping any bits from the edges. Cook until the tomatoes split and are hot, about 3 minutes. Stir in the vinegar.

4. To serve, spoon the tomato compote over the scallops, and top them with the nut mixture. Sprinkle the entire dish with rosemary and serve.

SERVES 4

Chilled Calamari Salad with Lemon and Basil

Up until the mid-'60s, the Vatican designated Christmas Eve a "fast and abstain" day, which meant you could only eat one meal, and it couldn't be meat. So if you only had one meal, you'd better make it matter. Given that Italy is a peninsula, fish was readily available, and the idea of a Christmas Eve fish banquet evolved. (Frowning on such grandiosity, the church fathers spun it their own way: if you're serving a lot of fish, then serve seven fishes, in remembrance of the Seven Sacraments.)

1	pound squid, cut into ⅓-inch rings and the tentacles halved
½	small red onion, thinly sliced
8	cherry tomatoes, quartered
¼	cup Kalamata olives, pitted and quartered
1	tablespoon capers
¼	cup extra-virgin olive oil
1	teaspoon lemon zest
1	tablespoon lemon juice
1	garlic clove, minced
2	tablespoons chopped fresh basil
¼	teaspoon salt
¼	teaspoon black pepper
	Pinch of red pepper flakes

1. Rinse the squid. Bring a pot of salted water to a boil, add the squid, and cook until the squid is opaque, about 50 seconds; do not overcook. Drain the squid and plunge it into ice water. As soon as it's cool, drain and pat dry.

2. Combine the onion, tomatoes, olives, capers, oil, lemon zest, and lemon juice, garlic, basil, salt, pepper, and pepper flakes in a bowl. Stir in the squid. Toss and refrigerate for 30 to 60 minutes before serving, to allow the flavors to blend.

SERVES 4

Grilled Squid *with* Lemon *and* Young Dandelion Greens

Like Elvis said, "love me tender." He might have added, make it quick. To keep squid tender, cook it quickly over high heat. If you want sustainable seafood, squid's for you.

1	pound squid, both bodies and tentacles
2	tablespoons lemon juice plus more to finish
1	teaspoon lemon zest
4	tablespoons extra-virgin olive oil
1	tablespoon chives, chopped
1	teaspoon fresh oregano, chopped
¼	teaspoon salt
1	teaspoon black pepper
8	cups dandelion greens (or substitute watercress) with tender stems

1. Cut the squid tubes along the side to open flat, and lightly score one side. Combine the lemon juice, lemon zest, olive oil, chives, oregano, pepper, and salt in a bowl and toss with the calamari. Marinate 1 hour in refrigerator, turning at least once.

2. Prepare a hot fire in a gas or charcoal grill, and oil a grill pan.

3. Place the squid in the grill pan and grill 1½ minutes per side on the tubes, and 2 minutes per side for the tentacles. Toss in a bowl with the greens and finish with lemon juice.

SERVES 4-6

Rhode Island–Style Calamari WITH Romesco Dipping Sauce

With the largest squid-fishing fleet on the eastern seaboard based in Point Judith, it's not surprising that Rhodies consider themselves squid connoisseurs, and they've even developed a Rhode Island–style squid (it has an appetizing ring to it), in which the squid is fried with pickled peppers. You've got to love Rhode Island and its food predilections — squid has become such a regional favorite that in 2013, a state representative from Warwick proposed to the state assembly that they consider making Rhode Island–style calamari the state appetizer.

Frozen squid is available year-round, but it's best to look at the box for the place of origin. Squid coming from China can be of varying quality; poor-quality squid can be tasteless and rubbery. It's worth finding fresh domestic squid in a fish market.

2	cups buttermilk		½	teaspoon black pepper
½	pound calamari, body cut into rings and tentacles reserved		2	teaspoons smoked paprika
			1	teaspoon garlic powder
¼	cup mild jarred pickled banana peppers, drained		½	teaspoon onion powder
			½	teaspoon or more cayenne pepper
¼	cup hot jarred pickled banana peppers, drained		1	teaspoon salt
				Peanut oil
½	cup all-purpose flour			Romesco Sauce (see page 161)
¾	cup cornmeal			

1. Put the buttermilk in a medium bowl. Add the calamari and the pickled mild and hot peppers. Marinate the mixture for 1 hour.

2. In a shallow dish, combine the flour, cornmeal, pepper, paprika, garlic powder, onion powder, cayenne, and salt. Drain the calamari and peppers, and dredge them in the flour mixture until coated, then put them into a colander and shake off the excess flour.

3. Fill a large, heavy pot with 1 inch of peanut oil. Heat the oil to 350°F (180°C) and fry the calamari and peppers in batches until golden, 2 to 3 minutes. Transfer to paper towels with a slotted spoon and serve with warmed romesco sauce for dipping.

SERVES 4

FISHY FRIDAYS

Growing up in the shadow of Irish Boston, I well remember the wretched rubbery fish sticks served in school cafeterias on Fridays. Urban legend circulated a theory that sounded like a plot from a Le Carré thriller: a powerful bishop in medieval times made a secret pact with the fishing industry, requiring millions of Catholics to abstain from eating meat on Fridays as sacrificial penance and changing global economics as a result. Even centuries later, McDonald's introduced the Filet-o-Fish sandwich because burger sales plummeted on Fridays.

The actual rule was that Catholics refrain from eating warm-blooded animals on Fridays in recognition of Jesus' sacrifice. With 4,700 miles of coastline, seafood was plentiful in Italy, and fish was more appealing than, say, eating snake. As the number of meatless days increased on the Christian calendar in medieval times (Fridays, Advent, Lent, holy days, etc.) the demand for fish grew, and became a powerful component in the growth of the global fishing industry. But it wasn't a papal plot — there was no snake in the grass.

Grilled Squid Devil Dogs with Habanero-Lime Ketchup

Also known as Point Judith Po' Boys, devil dogs are served at Smoke & Pickles in Westport, Massachusetts, near the Rhode Island border, when fresh whole squid is available locally in early May for a few weeks and then again in early November. The habanero-lime ketchup is tastiest when made a few days in advance. Can you tell that chef Marc DeRego used to work with *Thrill of the Grill* author Chris Schlesinger, of Inner Beauty hot sauce fame? This ketchup packs a punch. (It also makes a good dipping sauce for fried calamari.)

SQUID AND MARINADE

1½	pounds whole squid
2	tablespoons corn oil
1	tablespoon lime juice
1	tablespoon toasted coriander seeds, coarsely ground
⅛	teaspoon salt
	Freshly ground black pepper

HABANERO-LIME KETCHUP

½	large ripe red papaya, peeled, seeded, and diced
13	habanero peppers, cored and seeded
¼	cup firmly packed brown sugar
¼	cup white vinegar (plus more if a thinner sauce is desired)
1	cup yellow mustard
1	tablespoon chili powder
1	tablespoon curry powder
1	tablespoon cumin seeds, toasted in a dry skillet and ground

Top-split hot dog rolls
Butter
Crisp fresh lettuce

1. Clean the squid, then separate the tentacles and tubes and place in a nonreactive bowl.

2. To make the marinade: Put the oil, lime juice, coriander, salt, and pepper in a blender and blend on high until very smooth. Transfer the marinade to a bowl big enough to hold all of the squid.

3. To make the ketchup: Put the papaya, habanero peppers, brown sugar, vinegar, mustard, chili powder, curry powder, and cumin in the blender and blend until smooth.

4. Allow the squid to marinate lightly while you prepare your charcoal grill. Butter the sides of the hot dog rolls.

5. Grill the squid: The tentacles require about 3 minutes on the grill, turning as needed and taking care not to char the flesh. Grill the tubes for about 5 minutes.

6. Grill the rolls, and place the squid into the grilled rolls slathered with habanero-lime ketchup. Add lettuce and serve.

SERVES 4-6

205

SHAKESPEARE MAY HAVE SAID IT WAS "A VERY ANCIENT AND FISHLIKE SMELL," BUT WHEN IT COMES TO PREPARING FISH, IF YOUR FISH SMELLS FISHY, IT IS NOT FRESH.

Crab Cakes *with* Roasted Jalapeño Crema

Crab cakes are a personal thing; everyone has his or her favorite. These cakes are not coated in flour or crumbs and therefore have more crab flavor. Purchase the highest-quality crabmeat, preferably from a fish market on the shores of the Atlantic. The secret to keeping the cakes moist is a bit of bacon, or "fairy dust," as my cousin Donna calls it.

CRAB CAKES

2	slices bacon, diced
⅓	cup finely diced shallot or onion
⅓	cup finely diced red bell pepper
⅓	cup finely diced celery
1	egg
1	tablespoon heavy cream
1	teaspoon Dijon mustard
1	teaspoon seafood seasoning (such as Old Bay)
⅛	teaspoon cayenne pepper
½	teaspoon grated fresh ginger
½	teaspoon lemon zest
2	teaspoons minced flat-leaf parsley
½	cup panko breadcrumbs
1	pound jumbo lump crabmeat, picked over, shells and cartilage removed
	Neutral oil (sunflower or vegetable) for frying
2	tablespoons butter
	Sea salt

ROASTED JALAPEÑO CREMA

1	jalapeño pepper
1	cup ranch dressing
2	tablespoons chopped fresh cilantro
1	ripe avocado, peeled and pitted
	Juice of ½ lime
	Salt and freshly ground black pepper
6	lettuce leaves for plating
1	lemon cut in wedges, for serving

1. Place a skillet over medium heat and cook the bacon until the fat is rendered. Don't let the bacon get too crispy. Add the shallot, bell pepper, and celery, and sauté until softened, 4 minutes. Transfer the mixture to a bowl to cool.

2. Add the egg, cream, mustard, seafood seasoning, cayenne, ginger, lemon zest, and parsley to the cooled bacon mixture. Stir in the breadcrumbs and fold in the crabmeat, taking care not to break the large pieces. Refrigerate the mixture

for 15 minutes, then divide it into 12 patties of approximately 2 ounces each. Refrigerate again for 10 minutes to firm up the patties, which helps them hold their shape when frying.

3. To prepare the crema: Roast the jalapeño pepper under the broiler until blistered all over, about 2 minutes. Remove the core and seeds if you want less heat. Put half, or more, of the jalapeño and the ranch dressing, cilantro, avocado, lime juice, salt, and pepper in a food processor, and process until smooth. Strain the crema through a mesh sieve to obtain an even consistency. You might want to put it in a squeeze bottle for ease in drizzling over the crab cakes. Refrigerate. The crema can be made 1 week in advance.

4. Fill a large skillet with ½ inch of oil. Heat the oil until hot but not smoking, then add the butter. Gently lay the cakes in the pan, four at a time. Do not overcrowd the pan, or the cakes will steam instead of frying and not brown. Pan-fry until lightly golden and crisp, 4 minutes on each side. Drain on paper towels and season while hot with sea salt.

5. For each serving, place two cakes on a lettuce leaf to absorb excess oil, and top with a drizzle of crema. Serve with lemon wedges.

SERVES 6 AS AN APPETIZER

Crab Quesadilla WITH Charred Corn and Avocado Salsa

Be sure to use fresh crabmeat in this recipe. The salsa would be delicious with any grilled fish.

½	pound fresh crabmeat
½	cup grated mild cheddar cheese
	Freshly ground black pepper
4	soft flour tortillas
1	tablespoon grapeseed oil
	Charred Corn and Avocado Salsa (recipe follows)

1. Mix the crabmeat, cheese, and pepper in a small bowl. Lay the tortillas flat and place ½ cup of the crabmeat mixture on one side of each, spreading to within ½ inch of the edge. Fold the tortilla over.

2. Heat the oil in a large skillet over medium heat. Carefully place the quesadillas one at a time in the skillet and fry until browned, about 3 minutes; then flip and cook the other side, another 3 minutes, making sure the cheese is melted. Cut each one in half and serve with the salsa.

SERVES 4

Charred Corn Avocado Salsa

1½	cups fresh or frozen corn
1	small jalapeño, cut in half lengthwise and seeded
1	ripe Hass avocado, peeled, pitted, and diced
1	small tomato, diced
⅓	cup diced red onion
¼	cup chopped cilantro
	Juice of 1 lime
1	tablespoon olive oil
⅛	teaspoon salt
	Freshly ground black pepper

1. Preheat the broiler.

2. Lightly spray a baking sheet, then lay the corn and jalapeño on it in a single layer. Broil until the corn and jalapeño are charred, about 8 minutes.

3. Mince the jalapeño and put in a bowl with the corn. Add the avocado, tomato, onion, cilantro, lime juice, oil, salt, and pepper, and toss.

SERVES 4

Crab Empanadas

I fell in love with fish empanadas (a stuffed pastry that is baked or fried) after spending time in Vieques, Puerto Rico. This version is a handy way to use up leftover crabmeat; you could also substitute shrimp or fish.

DOUGH

1½	cups all-purpose flour
¾	teaspoon baking powder
¼	teaspoon salt
1	(8-ounce package) cream cheese, softened
4	tablespoons butter, softened
1	egg (for egg wash)

CRAB FILLING

1	tablespoon olive oil
1	roasted red pepper, diced
½	cup yellow onion, diced
2	teaspoons minced jalapeños
1	garlic clove, minced
½	pound crabmeat
2	tablespoons chopped basil
2	tablespoons chopped cilantro
¼	teaspoon salt
	Freshly ground black pepper

1. To make the dough: Sift the flour with baking powder and salt. In a large bowl, mix the cream cheese and butter with an electric mixer, adding the flour gradually and mixing until smooth. Cover and refrigerate for 30 minutes.

2. To make the filling: In a sauté pan, heat the oil and sauté the red pepper, onion, and jalapeños for 5 minutes. Add the garlic and sauté 1 minute more. Remove the pan from the heat, and add the crab, basil, cilantro, salt, and a few grindings of pepper.

3. Preheat the oven to 350°F (180°C).

4. To assemble the empanadas: Lightly flour a surface and divide the dough into four pieces (keeping it in the refrigerator as you use one piece at a time). With a floured rolling pin, roll out the dough ⅛ inch thick, and cut it into 4-inch circles. Put a tablespoon of filling in the center of each circle, fold the dough over to make a half circle, and crimp the edges. (You can make larger empanadas if you wish.) As you make the empanadas, cover them with a dish towel to keep them moist.

5. In a small bowl, make an egg wash by whisking the egg with a teaspoon of water. Assemble the empanadas on a lightly greased baking sheet, brush with the egg wash, and bake until golden, 15 to 20 minutes.

MAKES 16 EMPANADAS

Shrimp *and* Crab Manicotti *with* Cream Sauce

Unless you are making everything from scratch, this beloved Italian-American dish is an easy midweek meal. You can easily substitute other seafood or sausage.

14	manicotti shells
6	ounces lump crabmeat, flaked
6	ounces fresh peeled and deveined shrimp (see page 28), cooked and chopped
1½	cups shredded cheese (equal parts Parmesan, mozzarella, and Asiago)
1	cup part-skim ricotta cheese
½	cup chopped Italian parsley
2	eggs, beaten

CREAM SAUCE

1½	tablespoons butter
1½	tablespoons olive oil
2	tablespoons shallots, finely minced
2	garlic cloves, minced
3	tablespoons all-purpose flour
2	cups low-fat milk
¼	teaspoon Old Bay seasoning
¼	teaspoon salt
	Freshly ground black pepper
	Pinch of nutmeg

1. Bring a large pot of water to a boil and cook the manicotti according to the package directions. Drain and set aside.

2. Preheat the oven to 350°F (180°C).

3. In a medium bowl, combine the crabmeat, shrimp, 1 cup of the shredded cheese, ricotta cheese, the parsley, and the eggs. Fill the manicotti shells with the mixture and place in an ungreased baking dish large enough to hold them flat in one layer.

4. To make the cream sauce: Heat the butter and olive oil in a saucepan over medium heat. Add the shallots and garlic, and sauté until soft, about 3 minutes. Add the flour, stirring for 1 to 2 minutes to create a roux. Whisk in the milk and stir until thickened, 8 to 10 minutes. Remove from the heat and add the Old Bay, salt, pepper, and nutmeg. Taste the sauce and adjust the seasonings.

5. Pour the cream sauce over the stuffed manicotti shells and sprinkle with the remaining ½ cup of cheese. Bake, uncovered, for 30 minutes.

SERVES 6

BERT AND I

"I shot my dog the other day."

"Was it mad?"

"It weren't so damned pleased."

So goes the opening to *Bert and I*, an ode to the deadpan humor of the Pine Tree State. The first *Bert and I* album was created in 1958 by two students at Yale (one of whom had learned about Downeast Maine humor from a classmate at St. Paul's School) who spent a week in Maine, then cut 50 copies of a 10-inch vinyl record as a lark for friends and family. Sound effects were made by their mouths with a dorm room wastebasket serving as the echo chamber — the boat engine starting up is particularly memorable. Neither Marshall Dodge nor Bob Bryan had any idea that the perfect pitch of their amateur storytelling would take off, propelling them to performances at college campuses (a modern-day *Moth*!), several more albums, and a fan base that includes Garrison Keillor.

Throughout my childhood, my family repeated the *Bert and I* phrases often, applying them to all manner of circumstance.

"How do you get to (name a place)?" "You can't get there from here."

"Have you lived here your entire life?" "Not yet."

Or the Texan who said: "I don't know about your farm in Maine, mister, but I have a ranch in Texas that is so large that it takes me five days to drive around my entire spread." To which Bert replied, "I have a car just like that myself."

Prawns in Tomato Sauce with Feta

This dish uses ouzo, an anise-flavored aperitif that is popular in Greece. Use the largest shrimp available to you.

TOMATO SAUCE

1	medium red or green bell pepper, half seeded and quartered, half diced
1	medium onion, half quartered, half diced
2	large garlic cloves, 1 whole, 1 minced
4	tablespoons Greek or other olive oil
1	tablespoon tomato paste
1	(28-ounce) can peeled plum tomatoes, crushed
1	bay leaf
1	tablespoon sugar
¼	teaspoon crushed red pepper flakes (optional)
1	teaspoon dried Greek oregano
	Salt and freshly ground black pepper
½	cup cold water

12	jumbo shrimp or 1 pound U-12 shrimp, raw
3	ounces ouzo or Sambuca
4	ounces feta cheese, crumbled
1	lemon, halved
1	tablespoon chopped fresh mint
1	tablespoon chopped flat-leaf parsley
2	pita breads, toasted and halved just before serving

1. To prepare the tomato sauce: Put the quartered pepper, quartered onion, and whole garlic clove in a blender or food processor and pulse until finely minced. In a skillet, sauté the mixture in 2 tablespoons of the oil for 4 minutes, taking care not to let it brown. Add the tomato paste, crushed tomatoes, bay leaf, sugar, pepper flakes if using, oregano, salt, pepper, and water. Cover and cook over low heat, for about 90 minutes, stirring occasionally, until the sauce thickens.

2. Remove the shells from the shrimp, leaving the tails intact for presentation purposes. Devein by running a sharp knife down the back of the shrimp to remove the black streak and wash under cold water. Refrigerate until ready to use.

3. Preheat the broiler.

4. Heat the remaining 2 tablespoons of oil in a pan over medium heat, then add the diced pepper, diced onion, and minced garlic, and continue cooking until softened, 3 minutes. Add the ouzo and the tomato sauce, and cook until reduced, about 5 minutes. Add the shrimp and 2 ounces of the feta and simmer lightly for 5 minutes. When the shrimp is almost cooked through and has a pink, opaque color, remove the pan from the heat.

5. Divide the shrimp among four terra-cotta (traditional) or ceramic dishes, then add the sauce, and crumble the remaining 2 ounces feta over the top of each dish. Broil until bubbly, about 3 minutes.

6. Remove and let stand 2 minutes. Squeeze lemon juice over the top and garnish with mint and chopped parsley. Serve with toasted pita bread.

SERVES 4 AS AN APPETIZER

"I've given parties that have made Indian rajahs green with envy. I've had prima donnas break $10,000 engagements to come to my smallest dinners. When you were still playing button back in Ohio, I entertained on a cruising trip that was so much fun that I had to sink my yacht to make my guests go home." — F. SCOTT FITZGERALD

Marinated Shrimp *with* Fresh Herbs

This marinated shrimp is delicious on its own or with Mustard-Lime
Dipping Sauce (recipe follows).

16	jumbo shrimp
1	lemon, halved, with 3 strips lemon peel reserved
1	cup extra-virgin olive oil
2	shallots, thinly sliced
1	teaspoon red pepper flakes
1	garlic clove, grated
1	tablespoon chopped chives
1	tablespoon minced fresh oregano
1	tablespoon minced fresh thyme
¼	teaspoon salt
¼	teaspoon black pepper
1	lemon, cut into wedges, for serving

1. Using kitchen shears, score each shrimp along the back; this will make them easier to peel later.

2. Bring a medium pot of salted water to a boil. Squeeze in the lemon juice, add the shrimp, and cook until pink, 3 to 4 minutes. Cool in a bowl of ice water, drain, and peel.

3. Whisk the oil, shallots, pepper flakes, lemon peel, garlic, chives, oregano, thyme, salt, and pepper in a big bowl. Add the shrimp, toss to coat, and refrigerate for several hours. Serve cold with lemon wedges.

SERVES 4

NOTE

Cooking the shrimp with their shells on helps retain flavor.

You can cook shrimp a day ahead if you keep it covered and chilled.

Mustard-Lime Dipping Sauce

This dipping sauce goes well with marinated shrimp, grilled halibut, or salmon.

½	cup mayonnaise or sour cream
	Juice of 2 limes
2	tablespoons spicy mustard
2	teaspoons honey or brown sugar
½	teaspoon Worcestershire sauce
¼	teaspoon Tabasco or other hot sauce
½	teaspoon coarse salt
	Freshly ground black pepper

Whisk together in a small bowl the mayonnaise, lime juice, mustard, honey, Worcestershire, Tabasco, salt, and pepper. Chill until serving.

MAKES ½ CUP

218

A LOVE STORY

At the head of Narragansett Bay lies Providence, a terrific food city and melting pot, home to Brown University and Johnson & Wales. Waves of immigrants have flooded this city — Federal Hill used to be called Rhode Island's Little Italy, given its strong mafia and Italian presence.

Now it's touristy, catering to the college crowd, with Italian delis crowded between Lebanese restaurants and hookah bars. There remain many good Italian bakeries (never open on Sundays) in Providence and Cranston, where a lot of Italians relocated (remember Jenny Cavelleri from *Love Story*?).

Driving through Cranston in search of bakeries, I found tidy fenced yards with bright reflective balls and other religious ornaments, a long stretch of highway loaded with dry cleaners and Del's Lemonade stands, and a faded movie house with a lit-up neon sign "Welcome Back Engelbert Humperdink" (really?) and advertising the fact that Peter Noone and Herman's Hermits would be performing in a few months.

Indeed, the 1970 classic *Love Story* was in some regard a story of coastal New England, when a quick-witted daughter of an Italian baker from Cranston attends Radcliffe and meets a handsome Harvard blueblood from the North Shore.

OLIVER BARRETT IV AND JENNIFER CAVALLERI'S FIRST CONVERSATION:

"What makes you so sure I went to prep school?"

"You look stupid and rich," she said, removing her glasses.

"You're wrong," I protested. "I'm actually smart and poor."

"Oh, no, Preppie. I'm smart and poor."

She was staring straight at me. Her eyes were brown. Okay, maybe I look rich, but I wouldn't let some 'Cliffie — even one with pretty eyes — call me dumb.

"What the hell makes you so smart?" I asked.

"I wouldn't go for coffee with you," she answered.

"Listen — I wouldn't ask you."

"That," she replied, "is what makes you stupid."

Grilled Shrimp WITH Spicy Ginger Noodles

Also known as soba, buckwheat noodles (made from buckwheat flour) became popular in Japan during the late 1800s. Fat- and cholesterol-free, they're a good source of magnesium, lean protein, and thiamine.

8	ounces soba noodles (buckwheat)
2	teaspoons minced fresh ginger
2	garlic cloves, minced
2	tablespoons rice wine vinegar
2	tablespoons soy sauce
⅓	cup grapeseed oil plus 2 teaspoons
2	teaspoons sriracha sauce
2	ounces snow peas, julienned
4	scallions (green and white), thinly sliced
½	orange bell pepper, seeded and julienned
2	tablespoons toasted sesame seeds
16	medium shrimp, shells removed
	Salt and freshly ground black pepper

1. Bring a large pot of salted water to a boil. Cook the noodles until al dente, 2 to 3 minutes. Drain and rinse with cold water.

2. In a large bowl, combine the ginger, garlic, vinegar, soy sauce, ⅓ cup grapeseed oil, sriracha, snow peas, scallions, bell pepper, and sesame seeds. Add the noodles, mix well with your hands, and adjust the seasonings.

3. Devein the shrimp by running a sharp knife down the back of the shrimp to remove the black streak, and wash under cold water.

4. Prepare a cast-iron skillet by putting it over medium-high heat until it is very hot. Toss the shrimp with the 2 teaspoons grapeseed oil and salt and pepper to taste, and sear on both sides until cooked through, about 1 minute per side.

5. Divide the noodles among four bowls or plates, and top with the shrimp.

SERVES 4

Shrimp in Crazy Water

When I heard that *gamberoni all' acqua pazza* (Italian for "shrimp in crazy water") was on Mario Batali's menu for his last meal on earth, I had to play with it, especially given my penchant for spicy foods. The roots of this dish are in Naples, stemming from the days when fishermen would bring home their catch and sauté it in tomatoes, olive oil, and seawater. Traditionally made by poaching white fish in a broth, this recipe works well with lobster or any white fish (bass, cod, halibut, sea bass). While I've made it here with peeled shrimp, if you use jumbo shrimp with the heads and tails left on, it becomes a messy, roll-up-your-sleeves, eat-at-a-picnic-table sort of dish — part of the charm. The spicy broth begs to be served with Garlic Crostini (page 154) and Boston Salad with Blood Oranges and Red Onion (page 266).

4	tablespoons extra-virgin olive oil
1	medium onion, thinly sliced
4	garlic cloves, thinly sliced
½	fennel bulb, thinly sliced
1	(15-ounce) can diced tomatoes with juice
1	cup dry white wine
½	cup seawater or ½ cup water mixed with ½ teaspoon sea salt
½	cup chopped parsley
10	Kalamata olives, pitted and chopped
1	teaspoon cayenne pepper
1	pound large shrimp, peeled and deveined (see page 22)
	Freshly ground black pepper

1. Heat the oil in a large pot over medium heat. Add the onion, garlic, and fennel, and sauté until golden, about 8 minutes.

2. Add the tomatoes with their juice, the wine, and the seawater, and bring to a boil, then reduce heat to low and cook for 10 minutes. Add the parsley, olives, cayenne, and shrimp, cover, and cook until the shrimp is pink, 3 to 5 minutes. Season with black pepper (and more cayenne if desired) to taste. Ladle into individual bowls and serve.

SERVES 4

Roasted Tomatoes *with* Shrimp, Red Pepper, *and* Feta

No one talks about their food failures, right? I've had some doozies. My first "food" date (where I cooked for a guy), I was so nervous that I used 20 bulbs of garlic (rather than 20 cloves) to make Twenty Garlic Chicken. Once, on a boat trip, I mixed up powdered lemonade with water left over from boiling potatoes. (I was going to write that "I was so, so exhausted that I . . ." but the truth is: I just did it. It was a mistake.)

Cooking is all about building on information. You start with a peanut butter sandwich and go from there. Below is a really simple main course that can be served over couscous or pasta. It's a way to make winter tomatoes edible, and it's great with late-summer tomatoes, too. The shrimp offers a quick, lean protein.

6	medium tomatoes, cut into eighths
3	tablespoons olive oil
4	garlic cloves, minced
¾	teaspoon salt
¾	teaspoon black pepper
1	pound medium shrimp, peeled and deveined
1	cup feta cheese, crumbled
½	cup jarred roasted red peppers, sliced
2	tablespoons white wine
¼	cup chopped fresh basil
¼	cup chopped fresh parsley
	Juice of ½ small lemon (about 1 tablespoon)

1. Preheat the oven to 450°F (230°C) and place a rack in the top third of the oven.

2. Place the tomatoes in a large baking dish and toss them gently with the olive oil and garlic. Sprinkle with salt and pepper and roast on the top rack for 15 minutes.

3. Remove the tomatoes from the oven and stir in the shrimp, feta, red peppers, wine, basil, and parsley, then bake another 8 to 10 minutes or until the shrimp turn pink. Remove from the oven, squeeze the lemon juice over the dish, and serve hot or at room temperature.

SERVES 6

Skewered Tequila-Lime Shrimp with Mango-Jicama-Habanero Relish

Three ingredients: Tequila. Lime. Shrimp. These kabobs rock, especially when combined with mangoes, jicama, and habanero chiles. Jicama is a sweet root vegetable that tastes like an unripe pear. Metal skewers are fine, but if you use wooden ones, soak them for half an hour before grilling.

- ¼ cup gold tequila
- ¼ cup lime juice
- 2 tablespoons chopped cilantro
- ¼ teaspoon salt
  Freshly ground black pepper
- ¼ cup grapeseed oil
- 16 extra-large shrimp, peeled and deveined (see page 28)

MANGO-JICAMA-HABANERO RELISH

- ½ small jicama, peeled and diced
- 2 champagne mangoes, peeled, seeded, and diced
- ¼ cup diced red bell pepper
- 2 tablespoons lime juice
- 1 tablespoon grapeseed oil
- ¼ cup diced red onion
- ½ small habanero, seeded and finely minced
- ¼ teaspoon sea salt
- 2 teaspoons minced cilantro (optional)

1. To prepare the marinade: Combine the tequila, lime juice, cilantro, salt, pepper, and oil in a bowl. Add the shrimp and marinate, covered and refrigerated, for 30 to 60 minutes.

2. To make the relish: Combine the jicama, mangoes, bell pepper, lime juice, oil, onion, habanero, and sea salt, and let stand at room temperature for at least an hour.

3. Prepare a medium-hot fire in a gas or charcoal grill. Skewer the shrimp and grill 1½ minutes per side, or until cooked through, taking care not to overcook. Dust with cilantro, if desired, and serve with the fiery relish.

SERVES 2-4

Shrimp-Crusted Pork Loin *with* Chipotle Vinaigrette *and* Cassava Chips

Pairing shrimp with pork is common in many cuisines. Cassava (a root vegetable) is also known as yuca (yoo-kah), not to be mistaken with yucca, which comes from the root of a different plant. You can find yuca at some grocery stores, and often at Latin and Caribbean markets.

CASSAVA ROOT CHIPS WITH SPICY SALT

2   pounds cassava root, peeled and sliced paper thin
    Vegetable oil
1   tablespoon kosher salt
1   teaspoon garlic powder
1   teaspoon ground cumin
1   teaspoon cayenne, blackened spice, or taco seasoning

SHRIMP-CRUSTED PORK

4   (6-ounce) pork slices, cut from a boneless loin, ½ inch thick
    Salt and freshly ground pepper
1   tablespoon vegetable oil
2   tablespoons butter
1   tablespoon chopped garlic
1   tablespoon fresh cilantro leaves, chopped
¼   teaspoon red pepper flakes
1   pound shrimp (21–25 count), peeled and deveined, patted dry, and chopped
½   cup panko breadcrumbs

CHIPOTLE VINAIGRETTE

½   cup white wine vinegar
2   egg yolks
1   tablespoon chopped chipotle peppers in adobo sauce
1   tablespoon shallot, chopped
1   cup lightly packed fresh cilantro leaves, plus 1 tablespoon for garnish
1   cup olive oil

1. Place the sliced cassava in cold water to prevent browning.

2. To prepare the cassava chips: Heat 2 inches of vegetable oil (enough for deep frying) to 365°F (185°C) in a 2-quart pan. Dry the cassava slices on paper towels and fry them in the hot oil until slightly colored and crisp, 3 to 5 minutes. Combine the salt, garlic, cumin, and cayenne in a paper bag and toss with the fried chips. Store the chips in the bag until ready to use. They can be prepared one day in advance.

3. Preheat the oven to 350°F (180°C).

4. To prepare the pork: Pat dry the pork slices and season with salt and pepper (flour lightly, if desired). Preheat a heavy skillet until smoking, then add a tablespoon of oil to coat the pan. Sear the pork steaks on both sides, 3 to 4 minutes per side, and then transfer them to a baking sheet.

5. In the same pan over reduced heat, add the butter, garlic, the 1 tablespoon cilantro, and the pepper flakes. Sauté until the garlic is translucent. Add the shrimp and cook until slightly pink but not fully cooked. Add the breadcrumbs and stir. Remove the pan from the heat and divide the shrimp mixture among the pork slices by placing a spoonful on top of each. Bake for 10 to 15 minutes or until the pork reads 165°F (74°C) on a meat thermometer.

6. To make the vinaigrette: Put the vinegar, egg yolks, chipotle peppers, shallot, and cilantro in a food processor or blender and pulse several times to combine. Drizzle in the olive oil and blend to emulsify the dressing. Take care not to over-blend, or the oil will heat up from the motor, causing an unpleasant flavor.

7. Drizzle the pork with the vinaigrette and sprinkle with cilantro leaves. Serve with the cassava chips.

SERVES 4

Apricot-Glazed Coconut Shrimp with Pineapple-Tequila Salsa

This recipe provides a change in latitude. Sip a piña colada or tequila with lime and invite a few friends over for a cookout. While the pig is roasting, enjoy this crunchy shrimp appetizer that has a taste of the islands.

12	Colossal-size shrimp or 1 pound U-12 shrimp, raw
1	cup dried sweetened coconut
2	tablespoons flour
	Sea salt and freshly ground black pepper
2	egg whites
1	cup panko breadcrumbs
	Vegetable oil

APRICOT GLAZE

½	cup apricot preserves
2	ounces orange juice
	Pinch of cumin and salt
	Shake of Jump Up and Kiss Me or other hot sauce
	Pineapple-Tequila Salsa, for serving (recipe follows)
1	lime, cut in wedges, for garnish
	Cilantro leaves, for garnish

1. To prepare the shrimp: Remove the shells but leave the tails intact. Devein the shrimp by running a sharp knife down the back to remove the black streak and then washing the shrimp under cold water.

2. Preheat the oven to 200°F (100°C). Place the coconut on a baking sheet and dry it in the oven for 20 minutes. (Removing moisture ensures the coconut coating will fry to a crisp crunch rather than a chewy consistency.) Do not let the coconut brown; it should just dehydrate.

3. When ready to fry the shrimp, place the flour in bowl. Season the shrimp with salt and pepper to taste, and then toss them in the flour.

4. In another bowl, whip the egg whites until they reach meringue consistency and hold stiff peaks.

5. In a third bowl, mix the breadcrumbs and the dry coconut. Holding each shrimp by the tail, first dredge it in the meringue and then in the crumb mixture. Arrange the shrimp on a baking sheet.

6. Put ¼ inch of oil in a large, heavy skillet and heat to 350°F (180°C) over medium-high heat. Fry the shrimp until they float (meaning the inside is at temperature), approximately 3 minutes per batch. Do not overcrowd the pan. Drain the fried shrimp on a paper towel to remove excess oil. Lightly season them with sea salt.

7. To make the glaze: Put the preserves, juice, cumin, salt, and hot sauce in a small saucepan and cook over medium heat, stirring constantly, bringing the sauce to a simmer. Reduce by one-third, about 6 minutes. Keep warm.

8. Mound ¼ cup salsa on each of four plates. Top each serving with three crispy warm shrimp and drizzle glaze on each (or put the glaze in a bowl for dipping). Garnish with lime wedges and cilantro, and serve.

SERVES 4 AS AN APPETIZER

Pineapple-Tequila Salsa

- ½   medium ripe pineapple, julienned
- ¼   jalapeño pepper, minced (seeds removed if desired for less heat), optional
- ¼   green bell pepper, julienned
- ¼   red bell pepper, julienned
- ¼   cup chopped Bermuda or red onion
- 1   orange, divided into segments
- 1   tablespoon chopped cilantro
-     Splash of tequila
-     Sea salt and freshly ground black pepper

Combine the pineapple, jalapeño, green and red bell peppers, onion, orange, cilantro, tequila, and salt and pepper to taste in a small bowl. Stir well.

LOBSTER PRIMER

Most Yankees cook their lobsters by dunking them head-first in boiling water or tossing them on beds of seaweed and saltwater to steam over an open fire on a beach. Though Maine is famous for its lobster, there are lobsters in Buzzards Bay and off the Vineyard. The closer to shore, the smaller the lobster. Maine lobsters tend to be smaller than Massachusetts lobsters because the water gets deep very fast offshore.

Purchase lobsters the day you want to cook them. They should be alive and feisty. Keep them cool, stored on ice or in the refrigerator; not smothered in plastic wrap. Don't store lobsters in fresh water. Lobsters are rich in amino acids, potassium and magnesium, vitamin B_{12}, B_6, B_3, and B_2, calcium, phosphorous, iron, zinc, and vitamin A.

You have several choices for cooking lobster.

BOILING: Fill a large pot three-quarters full of seawater (or water with 2 tablespoons of salt per quart of water). You need about 2½ quarts of water per lobster. Bring to a brisk boil, then plunge one lobster in head-first. (Let the water return to a boil before adding the next lobster.) Lower the heat, cover, and simmer 15 minutes for a 1- to 1½-pound hard-shell lobster and 20 minutes for 1½- to 2-pound lobsters. Reduce cooking time by 3 minutes for new-shell (softer shell) lobsters.

STEAMING: This is the best method when you're cooking just a few. Put 2 inches of salted water in a large pot, or cook in salt water. When it's steaming, drop them in one at a time (holding them behind the claws) and cook for 18 minutes (1–1½ pounds) or 20 minutes (for 1½ pounds and up), reducing the time by 3 minutes for new-shell (softer shell) lobsters.

Some people judge a lobster to be done when it's red all over, but you can also tell by squeezing it — it's softer when cooked. A lobster that's just turned red may need a little longer, so the squeeze test is more reliable. Some people think an antenna will pull off when a lobster's done, but that doesn't always work.

Drain cooked lobsters on their backs and serve them hot, with melted butter.

Lobster Salad

Serve this salad on a bed of Bibb lettuce with some chopped cherry tomatoes on the side. It goes well with lavender lemonade (see page 272).

1	cup cooked lobster meat (from a 1–1¼-pound lobster)
3	tablespoons mayonnaise
1	teaspoon lemon juice
½	teaspoon chives
1	teaspoon celery seed
	Pinch of salt
1	small celery stalk, diced
⅛	teaspoon cayenne pepper

1. Chop the lobster meat into bite-size pieces.

2. Combine the mayonnaise, lemon juice, chives, celery seed, and salt in a bowl and stir to combine. Toss in the lobster meat and celery, and stir. Divide evenly between two plates and dust with cayenne.

SERVES 2

Let 'em Roll

This is an old-fashioned lobster recipe, typical lobster shack food. You might serve this with a chilled dry rosé when your great-aunt comes to visit. You can gussy it up by giving the sandwich a lettuce ruffle.

12	ounces cooked lobster meat, chopped
4	tablespoons finely chopped celery
2	teaspoons lemon juice
4	tablespoons mayonnaise
4	New England–style hot dog rolls
¼	teaspoon cayenne pepper
	Butter

1. Toss the lobster with the celery, lemon juice, and mayonnaise in a small bowl. Refrigerate until ready to serve.

2. Butter the outside of the rolls. Heat a skillet over medium heat and toast the rolls until golden, 1 to 2 minutes per side. Spoon the lobster salad into the rolls, and dust with cayenne pepper. Serve immediately.

SERVES 4

Lobster BLT

It's thought that the fourth Earl of Sandwich is responsible for the invention of the BLT. In 1762, in the middle of a gambling marathon, he asked his cook to prepare food that wouldn't interfere with the game. The cook served sliced meat between two pieces of toast — something that didn't require utensils and could be eaten one-handed.

1	tablespoon mayonnaise
¼	teaspoon lemon juice
¼	teaspoon fresh chopped chives
1	(1¼-pound) lobster, steamed and meat removed
4	slices good whole-grain bread
6	slices bacon, cooked crisp
1	small tomato, sliced
2	leaves Bibb lettuce

1. Whisk the mayonnaise, lemon juice, and chives in a small bowl.

2. Chop the lobster into bite-size pieces and toss with the dressing.

3. Toast the bread and assemble the two sandwiches by spreading the lobster salad on a slice of bread, followed by the bacon, tomato, and lettuce. Cut into halves and serve.

MAKES 2 SANDWICHES

WATERFRONT

Summering in New England as a kid, even then I was struck by how many rivers, estuaries, creeks, and tidal pools there were to explore. The beach town we lived in (Wareham) had over 54 miles of waterfront alone.

Miles of waterfront by state along this craggy shoreline:

- Maine 3,478
- Massachusetts 1,519
- Connecticut 618
- Rhode Island 384
- New Hampshire 131

Homemade Mayonnaise

Homemade mayonnaise is easy to make and tastes brighter than store-bought. You can vary mayonnaise with so many ingredients — harissa, hot sauce, anchovies, smoked paprika, minced garlic, red pepper flakes, sea salt, freshly ground black pepper. . . .

1	fresh or pasteurized egg (room temperature)
½	teaspoon dry or Dijon mustard
	Pinch of salt
2	tablespoons lemon juice
½	cup grapeseed oil
½	cup extra-virgin olive oil

1. Whisk together the egg, mustard, salt, and lemon juice in a small bowl.

2. Combine the grapeseed and olive oils in a measuring cup. While whisking the egg mixture, drizzle the oil into the egg mixture (in drops) until you see the mayonnaise form. Continue whisking vigorously (or use a food processor), adding a steady stream of oil.

3. Mayonnaise will keep, stored in the refrigerator, for 3 to 4 days.

MAKES 1 CUP

235

GETTING SAUCED

A 10th-generation Yankee, I grew up in a household where the only hot sauce was an old bottle of graying Tabasco that lived in the bar next to the vodka, where it was brought out judiciously on special occasions for Bloody Marys. It wasn't until my first boat delivery to the West Indies that I doused my fritters with an innocuous yellow sauce in a recycled ketchup bottle and immediately felt a freight train thundering through my sinus cavity. Once I recovered, I embraced the cult of hot. The heat. The humor. The flavor. The endorphin rush. The passion. Can you imagine an olive oil called Religious Experience? Mustard named Inner Beauty? Ketchup known as Last Rights? With as many hot sauce recipes as there are households, islanders taught me a thing or two about the fact that anything salt can do, hot sauce can do better. Stir it into a pot of boiling rice. Dab it on your oysters. Mix it with mayonnaise in a lobster sandwich.

Curried Lobster Roll

The bun is key: don't let anyone talk you into anything but a New England, top-split hot dog roll. Serve with Coleslaw with Carrots and Currants (page 268).

1	(1–1¼-pound) whole lobster, cooked (about 1 cup lobster meat)
3	tablespoons mayonnaise
2	teaspoons lime juice
½	teaspoon curry powder
2	teaspoons butter
2	hot dog rolls
2	leaves Bibb lettuce (optional)

1. Chop the lobster meat into bite-size pieces.

2. Combine the lobster, enough mayonnaise to coat, lime juice, and curry powder in a bowl and stir to combine.

3. Heat a skillet over medium heat, melt the butter, and toast the rolls until golden, 1 to 2 minutes per side. Transfer the rolls to serving plates, place a leaf of lettuce into the crook of each bun, and divide the lobster mixture between the two. Serve immediately.

SERVES 2

Homemade Curry Powder

Curry powders are easy to make and better in flavor and aroma than what you'll find in stores.

4	tablespoons coriander seeds
4	teaspoons cumin seeds
1	teaspoon cardamom seeds
1	teaspoon black peppercorns
1	teaspoon ground turmeric
1	teaspoon pure chile powder

Dry-roast the coriander, cumin, and cardamom seeds in a skillet until lightly browned and aromatic, about 1 minute, then grind in a spice grinder with the peppercorns. Combine with the turmeric and chile powder, shaking the curry spices together in an airtight container.

ONE MORNING IN MAINE

When I was quite young, we spent several years in a small winterized cottage in Falmouth Foreside, Maine. My mother read the Robert McCloskey books to me, and it felt like our life. In the mornings, when it was calm and flat, you could see porpoises playing in Mussel Cove. The path to the cove was laden with lady's slippers. At low tide, seals would sun themselves on rocks out on the point. On weekends when it breezed up, we'd sail over to Handy's Boat Yard in Falmouth for jelly doughnuts and to look at the tank filled with lobsters the fishermen had brought in. In a neighbor's Beetle Cat we'd sail to the islands in Casco Bay. I can still smell and hear the crinkle of dry pine needles under my bare feet on warm summer afternoons.

We eventually brought our big sailboat down Maine, and once when we were sailing, a storm blew up. Though we tacked for home there's no hurrying a sailboat, and we got caught in it. My mother got nervous, and stuck me in the companionway, where I sucked my thumb, held on, and watched them.

Nervous as my father reefed the mainsail while the winds howled and the flapping sails made a racket, my mother asked my father what she should do with the rope on the boat. "There *is* no rope on a boat," he snapped. It was the only time he ever got really fussy.

Stuffed Maine Lobster Tail Scampi

Lobster hasn't always been the luxury item it is today. During colonial times, lobster was plentiful. Native American women dove into the ocean to catch them to use the meat as bait for more desirable fish. Times have changed.

4	(8-ounce) uncooked Maine lobster tails
	Salt and freshly ground black pepper

STUFFING

4	tablespoons unsalted butter
1	large garlic clove, minced
1	medium onion, finely chopped (about 1 cup)
¼	cup finely chopped red bell pepper
1	egg
¼	cup mayonnaise
1	tablespoon Dijon mustard
2	teaspoons Worcestershire sauce
1	teaspoon seafood seasoning (such as Old Bay)
1½	cups crushed round butter crackers (such as Ritz)
1½	cups jumbo lump crabmeat, picked over, shells and cartilage removed

SCAMPI SAUCE

4	tablespoons unsalted butter, cold
2	tablespoons olive oil
1	large garlic clove, minced
¼	cup white wine
½	cup chicken broth
	Juice of 1 lemon
1	tablespoon chopped flat-leaf parsley
	Lemon wedges

1. Hold the lobster tail in your hand, with the hard, rounded shell upward and the tail flippers facing away from your body. Using kitchen shears, cut the shell up the back, toward the tail flippers, creating one long slit but keeping the tail intact. Squeeze the shell on both sides to expose the meat. Gently pull the meat upward, taking care to leave it attached and in the shell near the tail flippers. Devein by removing the intestinal tract from the meat, and somewhat "butterfly" the tail meat to expose more surface area. Keep refrigerated while preparing the stuffing.

2. Preheat the oven to 375°F (190°C) and set the rack in the upper third position.

3. To prepare the stuffing: Melt the butter in a skillet over medium heat. Add the garlic, onion, and bell pepper, and sauté until soft, about 3 minutes. Season with salt and pepper to taste, and let cool.

4. After the vegetables have cooled, stir in the egg, mayonnaise, mustard, Worcestershire sauce, and seafood seasoning. Stir in the cracker crumbs, and gently fold in the lump crabmeat, trying not to break up the large pieces. Season the lobster meat with salt and pepper. Divide the stuffing among the four tails, mounding it on top of the lobster meat on the back of the tail shell.

5. Bake the stuffed lobster tails until the stuffing is crisp and golden, 15 to 20 minutes, or until the lobster meat reaches 140°F (60°C) on an instant-read thermometer. If more browning is desired, broil for 2 minutes.

6. While the lobster is cooking, make the scampi sauce: In a skillet over medium heat, melt 2 tablespoons of the butter with the olive oil, add the garlic, and cook for 1 minute. Add the wine and simmer to reduce the liquid, about 3 minutes. Add the chicken broth and lemon juice. Season with salt and pepper and simmer for another 6 minutes. Whisk in the remaining 2 tablespoons cold butter to thicken the sauce.

7. When ready to serve, sprinkle the parsley over the sauce, then spoon the sauce over the cooked lobster tails. Serve with lemon wedges.

SERVES 4

MOUNT DESERT

A few years ago, my family went to Mount Desert Island. We enjoyed the famous haunts — hiking Bubble Rock and sailing on Somes Sound — and we ate at Thurston's, a lobster shack on stilts over the tidal pools in Bernard. Customers stand in line for corn and coleslaw, and then pick a lobster out of the tank; it arrives 10 minutes later on a red plastic tray. We all ordered, and our 10-year-old son Trainer asked if he could use his allowance to buy a lobster (that should have been a dead giveaway). Then he asked the waitress for it cold and uncooked, went outside, walked down the stairs, and set it free.

PORTLAND, MAINE

Portland's a no-tablecloth kind of place, with an animated restaurant scene that has turned it into one of America's best cities for foodies. With a spirited food culture (not to mention a vibrant music scene), those who live in this progressive port city (small enough that most of the chefs, artisanal food makers, and grateful customers know each other) revel in the back stories: turnip farmers, divers who deliver hand-harvested scallops to back doors, hot sauce makers, renegade bakers, and lobster guys who recount tales from ocean to table.

It's a forgiving city, too; unlike New York or D.C., there's a low barrier to entry, where a chef can set up a restaurant on the cheap and — without inherited European traditions casting long shadows — make a few mistakes and let culinary experiments ferment. History is present, yet there's a freshness in the salt air, a ruggedness that comes at least in small part from people who tough it out in a cold climate, plus a "culinary idealism and anarchy," according to the *New York Times*, where folks tuck in for a cold winter at places like Gritty McDuff's brewpub, which boasts beers such as "Punch You in the IPA," named after the Phish song "Punch You in the Eye."

There's also an awareness of sustainable seafood practices in this state with 3,478 miles of coastline, where for centuries people have depended on the sea for food, employment, and a way of life. Barton Seaver, who left a restaurant career in D.C. and moved to South Freeport to focus on sustainability, has a hefty title or three: not only is he a National Geographic Fellow, but director of the Healthy and Sustainable Food Program at the Harvard School of Public Health, and the New England Aquarium's first Sustainability Fellow. He thinks a key to seafood sustainability is to be pragmatic and encourage people to try lesser-known species, which means sharing stories behind the fish. "People want to be connected back to the fisherman," he notes.

Maine may be the ideal place to practice this gospel. "There is a can-do mentality and a fortitude of spirit that I've witnessed," he told one reporter. "I understand that I am saying these words as someone from 'away,' but that's part of what drew me here . . . people don't care what you know about until you show what you care about."

Shrimp *and* Lobster Ravioli *with* Pesto Cream Sauce

This is stupidly easy, as my friend Jody says. If you don't want to make your own pesto, there are good jarred or refrigerated pesto sauces on the market.

2¼	cups heavy cream
2	tablespoons pesto
8	ounces fresh lobster meat
8	ounces shrimp, peeled and deveined
2	tablespoons minced shallot
½	teaspoon salt
¼	teaspoon black pepper
1	(12-ounce) package round wonton wrappers (30 wrappers)
	Cornstarch
	Freshly grated Pecorino Romano cheese

1. Put 2 cups of the cream in a deep saucepan and bring to a simmer over medium heat. Simmer until it reduces by half, about 30 minutes. Remove from the heat, stir in the pesto, and cook for an additional 5 minutes. Keep warm.

2. Put the lobster, shrimp, the remaining ¼ cup cream, shallot, salt, and pepper in a food processor and process until minced.

3. Bring a large pot of salted water to a boil.

4. To make the ravioli, place 2 teaspoons of the seafood filling into the center of a wonton wrapper, moisten the edges with water, fold the wonton in half, and pinch the edges to crimp shut. As you finish the ravioli, lay them out on a baking sheet and sprinkle with cornstarch.

5. When all the ravioli are assembled, gently drop them into the boiling water. Cook over medium heat for 3 minutes, then remove with a slotted spoon. Place into individual bowls and top with cream sauce and a bit of cheese.

SERVES 6

> NOTE
>
> These ravioli freeze easily. Before cooking, while still on the baking sheet, cover lightly with plastic wrap and freeze. When solid, they can be transferred to a Tupperware container.

ON THE

BEACH

AN YOU THINK OF A PLACE you'd rather spend the day than on the beach? Well, maybe the water. Hunting for shells, swimming, body surfing, talking, sleeping, sunning, reading, eating, watching the horizon, discovering tidal pools — it's a day of relaxation and resourcefulness. While the kids run off looking for horseshoe crabs and shells, the adults take care of the rest.

Getting to the beach can be half the fun, whether it's driving to the Cape, taking a ferry to the Vineyard, paddling a kayak to a cove, paddleboarding to an island, or sailing across a bay. Sometimes loud and

surf-starchy, sometimes sheltered and remote, beaches beckon to be explored. Sea lavender. Rose hips. Rustling eelgrass. Fragrant pine groves. Shell beaches, rocky beaches, black beaches, white powder sand beaches. Whether you're setting anchor and rowing ashore, or taking your bike down to the beach, or driving with a surfboard in the back, the beach beckons, and a picnic is an integral part of experiencing the day.

Load up a daypack, a cooler, the tailgate, or your bike panniers with picnic foods. Think of foods that travel well; stay away from sandwiches that get soggy and salads that wilt. Foods that taste good at room temperature are always welcome on picnics. Maybe bring an insulated bottle of tea or good coffee (not to mention a blanket and sweatshirts) for late afternoon lounging as the shadows lengthen. Or consider an evening picnic, with blankets, a bonfire, and good sipping rum for enjoying the sunset before heading home. A picnic can be as simple as sandwiches on waxed paper or as elaborate as a complete meal served with cloth napkins, real china, crystal glasses, and candle lanterns. This chapter also includes instructions for a fabulous all-out beach day: an old-fashioned clambake. The real thing.

Lobster-Clambake

To New Englanders, the clambake is as cherished and steeped in tradition as Paul Revere's ride. Clambakes are primal rituals, big boisterous affairs, usually held in high summer. Like a tailgate picnic, they are best done with a group, for they take a lot of time, but lots of helping hands add to the fun. The hardest part is collecting wet seaweed. It is so worth it — don't be deterred.

Look at the tides, and make sure you build your bake at the high-water mark, as you'll need about 7 hours from start to the time you eat. After spending the day gathering seaweed and firewood, building the pit, and eating your feast, you'll want to linger by the fire to enjoy the night sky, so bring blankets, fixings for s'mores, and some good rum. Clambakes comprise three of my favorite activities: digging in the sand, enjoying a beach bonfire, and eating buttery shellfish with my fingers. This is the original Native American finger food — no bibs needed and you can wash your fingers in the surf.

FOR THE FEAST (SERVES 8–10)

- **8–10** lobsters, kept alive under wet seaweed
- **8** pounds steamers (soft-shell clams), soaked in seawater for several hours, drained, and put in a wire basket or tied up in several layers of damp cheesecloth
- **8–10** large potatoes, scrubbed and tied in several layers of damp cheesecloth
- **6** medium onions, loose skin removed, with an X cut in the root ends, and tied in several layers of damp cheesecloth
- **10** ears unhusked corn, soaked in seawater and tied in several layers of damp cheesecloth
- **3** pounds of spicy sausage (linquiça, chouriço, andouille, kielbasa), tied in several layers of damp cheesecloth
- **3** sticks unsalted butter

FOR THE FIRE

Several yards of cheesecloth or wire baskets to hold the clams, corn, onions, potatoes, and sausage

Canvas tarp (at least 5 feet square when doubled over)

Trash can or other container large enough to hold the canvas covered in seawater

Buckets for collecting seaweed

Seaweed (lots of it!)

Shovels for digging (the more the merrier — it's the pits to get stuck digging alone)

Firewood

Large stones that haven't been heated before

Iron rake, pitchfork, or boat hook

Heavy pan or saucepan with lid, for melting the butter

continued on next page

Lobster-Clambake Step-by-Step continued

 1. Divide and conquer!
Put the canvas in the trash
can, cover with seawater,
and leave to soak. Set a
group to digging a big
hole in the sand near the
high-water mark, 2 feet
deep by 4 feet wide. Send
another crew out to collect
enough seaweed to make
three heavy layers over
the fire. Dried seaweed can
be added to the trash can
to revive it.

2. Line the bottom and sides
of the hole with beach
stones the size of footballs.

3. Light a big hardwood
fire in the pit and
let it burn down.

4. Place more stones on the fire, making sure you don't smother it. Build another fire on top of the second layer of rocks and let it burn down.

5. Place the food in wire baskets or tie it up in loose cheesecloth slings. Dampen the cheesecloth with seawater so it doesn't catch on fire. You're now ready to cook.

6. Working quickly, rake the embers away from the rocks, then throw a full 6 inches of wet seaweed over the rocks.

continued on next page

Lobster-Clambake Step-by-Step continued

7

8

9

7. Put the potatoes, onions, and sausage on the seaweed and cover them with more wet seaweed. Add the lobster, corn, and clams. Cover with a final layer of wet seaweed.

8. Cover the mound with the wet canvas.

9. Seal the edges with sand and more seaweed so steam can't escape. Cook for 1½ hours, occasionally dousing the canvas with water. In the last 30 minutes or so, set the butter in a covered pan near the fire to melt.

10. When it's time to eat, dig the sand and seaweed away from the pit.

11. Carefully lift the corners of the canvas and peel it back, making sure sand doesn't get on the food.

12. Rake away the layers of seaweed, lifting out the food as it's exposed. Serve immediately with melted butter.

SERVES 8-10

WHAT GOES WELL WITH A CLAMBAKE?

- Coleslaw and other raw vegetable salads
- Easy appetizers to munch on, such as guacamole and chips, cheese and crackers, olives and nuts
- Beer, wine, and a good rum or whiskey for sipping on the rocks as the fire burns down
- Plenty of water, lemonade, and other thirst quenchers
- Watermelon or fresh mangoes for dessert
- Frisbee, kite, beach ball, volleyball, bocce ball, fireworks

Grilled Tuna Steaks _with_ Kimchi

Tucked in Maine's Penobscot Bay are two islands — North Haven and Vinalhaven — which, while twin islands and true havens, are worlds apart. North Haven is a prominent summer community settled by Boston Brahmins and Harvard types, while Vinalhaven is proud of its working harbor for lobstermen and draggers. They are separated by the winding Fox Island Thorofare, a popular spot for sailors and fishermen, who have been known to catch dandy blackfin tuna in these straits.

Traditional kimchi is a spicy, fermented cabbage dish. This version is pickled with a little vinegar and develops more quickly than lacto-fermented kimchi. It's ideal for a dinner with friends where you want to do advance prep work a few days ahead, or to take with you on your boat in anticipation of the fish you'll grill off the stern.

KIMCHI

1	(2-pound) head napa or bok choy cabbage
8	ounces daikon (a Japanese radish), peeled and cut into matchsticks
2	tablespoons sea salt (or other salt that is free of iodine)
3	tablespoons rice wine vinegar
2	teaspoons dark sesame oil
1	teaspoon grated ginger
5	garlic cloves, minced
1	teaspoon sugar or honey
1	teaspoon crushed Szechwan peppercorns
2	red jalapeños, seeded and julienned
4	scallions, trimmed and thinly sliced
4	boneless tuna steaks, 1-inch thick
1	tablespoon olive oil
	Salt and freshly ground black pepper

1. To prepare the kimchi: Trim the root end of the cabbage, and cut the cabbage into quarters lengthwise; then cut the quarters crosswise into 1-inch strips. Combine with the daikon and sea salt in a bowl and mix well. Put a plate on top to weigh the vegetables down, and let them stand at room temperature for 1 to 2 hours. Rinse under cold water in a colander.

2. Meanwhile, mix the vinegar, sesame oil, ginger, garlic, sugar, and peppercorns in a large bowl. Add the jalapeños and scallions. Add the cabbage and daikon (with some moisture) and toss with your hands to combine with the dressing,

coating the vegetables thoroughly. Pack the mixture into a glass jar with a tight-fitting lid and press down on the kimchi so the brine covers the vegetables. Cover the jar tightly, and refrigerate for 3 or 4 days before serving. (This recipe will make 4 cups of kimchi.)

3. When you're ready to grill the tuna, spray the grill grate with nonstick spray and prepare a hot fire in a gas or charcoal grill. Brush the tuna steaks with olive oil and season with salt and pepper on both sides. Grill the steaks 2 minutes on one side for medium-rare (1 minute for rare), then flip and grill an additional minute, or until desired doneness. (There should remain a pink translucence in the center.) Serve immediately with the kimchi on the side.

SERVES 4

NORTH HAVEN DINGHY

In 1885, William Weld (grandfather of the former Massachusetts governor William F. Weld and one of the founders of the North Haven summer colony in Maine) challenged other North Haven yachtsmen to a race in his tender (the dinghy from his 102-foot schooner yacht *Gitana*). He lost, went home, and that winter had a better dinghy designed and built, in which he sailed and beat everyone the next summer. At the end of the season, two copies were built by another summer resident. In 1888, four more dinghies were built, and a sailing fleet was born. The fleet grew, and today the North Haven dinghy is the oldest active racing class in the United States.

Whole Grilled Mackerel WITH Lime AND Spicy Yogurt

If you're having a beach picnic and throw out a line, you might just catch a fish — mackerel or not. And if you bring wood for a bonfire, you might just cook it up. This is a good recipe for smaller fish, whether you're at the beach or at home. Grilling with the skin on (which includes a thin layer of fat) helps insulate the meat, keeping it moist and tender. Other good choices here are snapper, flounder, or striped bass. "We're Atlantic ranchers," a lifeguarding friend reminisced to me recently. When we were teenagers, we'd fish for flounder, stripers, and blues (bluefish), catch blue crabs and lobsters, and dig for clams and quahogs during the summer. Other than vegetables and bread, in the warm months, pretty much everything came from the sea.

SPICY YOGURT

1	cup plain Greek yogurt
1	medium cucumber, peeled, seeded, grated, then lightly squeezed
1	tablespoon fresh lemon juice
½	teaspoon local honey
1	tablespoon chopped fresh cilantro
½	teaspoon salt
	Freshly ground black pepper
¼	teaspoon ground cumin
¼	teaspoon ground coriander
¼	teaspoon ground ginger
¼	teaspoon ground turmeric
1	garlic clove, minced
	Pinch of cayenne

FISH

2	whole fish (12 ounces to 1 pound each), cleaned, heads and tails intact
¼	cup olive oil plus more as needed
1	lemon, thinly sliced
1	tablespoon chopped basil
1	tablespoon chopped cilantro
	Sea salt and freshly ground black pepper
	Kitchen string, soaked
1	lime, sliced
	Cilantro and basil leaves, to garnish

1. To prepare the yogurt: Stir together the yogurt, cucumber, lemon juice, honey, cilantro, salt, pepper, cumin, coriander, ginger, turmeric, garlic, and cayenne.

2. Prepare a medium-hot fire in a wood or charcoal grill. Score the fish by cutting a slash lengthwise, then season each fish cavity with a drizzling of olive oil, lemon slices, basil, cilantro, and salt and pepper to taste. Tie each fish closed with the string.

continued on page 257

"They fried the fish with bacon and were astonished; for no fish had ever seemed so delicious before. They did not know that the quicker a fish is on the fire after he is caught the better he is; and they reflected little upon what sauce open air sleeping, open air exercise, bathing, and a large ingredient of hunger makes, too."

— MARK TWAIN

Whole Grilled Mackerel *with* Lime *and* Spicy Yogurt continued

3. Make several ¼-inch-deep diagonal slashes on both sides of the fish so they cook evenly throughout. Rub both sides of each fish with olive oil, season with salt and pepper, and work the oil into the slashes as well as the tail, which is wonderfully crisp when grilled.

4. Oil the grill grate and place the fish on the grill. Cook uncovered, until the underside is crisp and charred and the flesh along the gills on the underside is opaque, 5 to 10 minutes, depending on the size of fish. Gently flip each fish (two spatulas help) and grill on the other side until the skin is charred and the fish is flaky and opaque, 5 to 10 minutes. (When a small knife slides through the thickest part of the flesh, it's done.)

5. Place both fish on a serving platter, squeeze a little lime juice over the top, and garnish with lime slices and cilantro and basil leaves. (To eat, cut the fillet free from the top side of the fish and remove with a spatula. Remove the bones to free the other fillet.)

SERVES 2

257

LABOR DAY PICNIC

I think Labor Day weekend by the beach is a glorious end-of-summer ritual, and so was intrigued to learn that in Westport Point, Massachusetts, the Dexter family has maintained a Labor Day picnic tradition for well over a hundred years. Every year, interrupted only by wars and hurricanes, the Smith Dexter clan has hosted a picnic for family, friends, and neighbors. We're talking big here: up to 100 people attend the party. The centerpiece is a chowder that the men slow cook over a wood fire according to an old family recipe. "The sweet aroma of the simmering brew," wrote Gay Gillespie, "combined with wood smoke from an open fire is instantly recognizable to anyone lucky enough to attend. That signature aroma, combined with the smells of salt air, low tide, and the first autumn leaves, can only mean the Labor Day Picnic."

That's not all. For years, the picnic was held on a big rock in the East River. Guests were expected to motor, row, or sail out, bringing their own drinks and sandwiches to the picnic. By early morning, people in their nearby cottages were assessing the tides, the weather, and their picnic hampers. If you weren't on potato-peeling duty, you were out on the rock by midmorning to cook, while the kids clambered around — swimming, diving off the rock, mucking about in boats, clamming, and having fun. By noon, the feast began, with plenty of sherry ("in the old days they were pickled before lunch," Gay allowed), sandwiches, and chowder, the adults reminiscing about the summer, the kids squirreling about. By two o'clock, the rock would be deserted; everyone had to head back to shore to close up their cottages, put up the boats, and drive home with kids and pets. Another summer gone by.

Smoked Salmon Sandwich

A sandwich with flaked, smoked fish and horseradish yogurt piled on good bread will up your game on a boat or picnic, especially with the Spicy Cucumber Salad on page 260.

4	ounces cold-smoked or hot-smoked salmon
¼	cup sour cream or yogurt
2	teaspoons fresh horseradish
4	slices good whole-grain bread
1	dill pickle, thinly sliced
	Fresh dill sprigs

1. Flake the salmon into a small bowl. Combine the sour cream and horseradish in another bowl. Toast the bread.

2. To assemble sandwiches, lay out the toast. Spread sour cream mixture on each slice of bread, then top two of the slices with salmon, pickle slices, and dill. Close the sandwiches, cut, and serve.

MAKES 2 SANDWICHES

Spicy Cucumber Salad

This is good on its own and terrific at a clambake. It pairs well with that cold grilled salmon you might bring along on a picnic.

1	tablespoon sesame seeds
1	large cucumber, thinly sliced
2	tablespoons chopped scallion
2	tablespoons rice wine vinegar
1	teaspoon sesame oil
¼	teaspoon chili oil
¼	teaspoon finely grated garlic
½	teaspoon honey
⅛	teaspoon salt
	Freshly ground black pepper

1. In a small, dry skillet, toast the sesame seeds for 1 minute, shaking them often.

2. Put the sliced cucumber in a small bowl. Add the scallions, vinegar, sesame oil, chili oil, garlic, honey, salt, and pepper, and toss until everything is well coated. Let stand for 30 minutes at room temperature and then serve.

SERVES 4

Asian Cucumber Salad

This is one of my easiest go-to salads to make ahead for a picnic, cookout, or day on the water.

1	tablespoon sesame seeds
1	large cucumber, thinly sliced
¼	red onion, thinly sliced
¼	cup rice wine vinegar
1	tablespoon minced cilantro
½	teaspoon sea salt
¾	teaspoon honey
¼	teaspoon red pepper flakes
	Freshly ground black pepper

1. In a small, dry skillet, toast the sesame seeds for 1 minute, shaking them often.

2. Toss the sliced cucumber together with the onion, vinegar, sesame seeds, cilantro, salt, honey, pepper flakes, and pepper to taste until everything is evenly coated. Put in a container with a tight-fitting lid and refrigerate for an hour before serving, or up to 5 days.

SERVES 4-6

PICNIC TIPS

- Chill water, beer, and other drinks before packing them in the cooler to get the most out of your refrigeration.
- If you're using a cooler, don't drain the ice water; it helps prevent the rest of the ice from melting.
- Bring a bucket along and keep drinks cold by plunging them in seawater. If you're on a boat, dangle them overboard in a net.
- Pack a snack bag stocked with quick energizers such as cheese, crackers, nuts, dried fruits, apples, raisins, trail mix, and raw vegetables. If the weather's good, don't let hunger keep you from lingering at the beach.

ANGLING FOR THE BEST SPOT

So where are the most alluring fishing spots in New England? If you don't agree, blame *The Boston Globe*; it was their idea.

- **Best Beginners' Striper Surf Casting in Maine:**
Half Mile Beach

- **Best Bass Fishing from the Shore in Massachusetts:**
Outer Cape Cod

- **Best Summer Flounder in Connecticut:**
Fort Trumbull State Park

- **Best Bluefish in Massachusetts:**
Plymouth Harbor Jetty

- **Best Stripers from Shore for Seasoned Anglers in Rhode Island:**
Sandy Point (north shore of Cuttyhunk)

- **Best Mackerel from a New Hampshire Bridge:**
Goat Island Bridge

Old-Fashioned Cucumber Salad

For 20 years, I've had the same recipe tester: Jody Fijal. Both culinary eccentrics, we cook so well together that sometimes we finish each other's sentences. I admire her skills, culinary pragmatism, and palate, and we often share food. I give her eggs or honey from our bees; she saves me fresh horseradish or pickled daylilies she's put up. She'll surprise me with a caviar spoon she's found, and I know I'll crack her up when I give her cocktail napkins that say, "After Monday and Tuesday even the calendar says WTF."

For years Jody's talked about her Aunt Wanda, a single woman, born in 1925, who left home to work as an executive secretary at IBM, liked to travel, and after retiring returned to the Berkshires to take care of her father and brother. This is Wanda's cucumber salad recipe.

3	**English cucumbers, peeled and thinly sliced**
1	**tablespoon kosher salt**
¾	**cup sour cream**
2	**tablespoons lemon juice**
2	**tablespoons fresh chopped dill**
	Salt and freshly ground black pepper

1. Put the cucumbers in a colander inside a bowl or the sink, and toss them with the kosher salt. Let stand for 1 hour.

2. Put the cucumbers into a clean linen dishcloth and squeeze out the juice. Put the dry cucumbers in a bowl, add the sour cream, lemon juice, dill, and salt and pepper to taste, and toss until well combined. Refrigerate the salad until chilled, at least 1 hour.

SERVES 8

Salad Greens *with* Great Hill Blue Cheese *and* Vinaigrette

Sixty miles south of Boston lies a peninsula of 300 acres of woods and meadows that juts out into Buzzards Bay. The highest point on the property is Great Hill, a glacial mound rising 124 feet above sea level. A sacred tribal meeting ground long before 29 Pilgrim families from Plymouth Plantation settled in Marion in 1678, Great Hill was where a peace accord was signed during King Philip's War by English troops and Queen Awashanks and her Sakonnet tribe.

The property has been owned since 1908 by the descendants of Galen Stone, a Bostonian who co-founded Hayden, Stone & Company in 1892. In 1911, he tore down the house on Great Hill and built a castle-like Tudor mansion with Hayden stone, which he ordered from a quary outside Philadelphia and had shipped on barges to Marion because he liked the name. It was an enormous summer "castle" with 30 servants in the main house. There were also 12 greenhouses (prize-winning orchids and papayas were grown), and a barn for the Guernsey cows.

Over the years many guests stayed at Great Hill, including my grandparents. I loved curling up in my grandmother's wing chair, listening to her stories about going out on the Stones' yacht *Arcadia* in the 1920s, where multi-course dinners would be served, starting with oysters in carved shallow crystal bowls that, on closer inspection, were made of ice. The Stone Estate, as it's called locally, is still privately owned, with many descendents of Galen Stone living on the property. Prize-winning Guernsey cows roam the property, and Great Hill Blue Cheese is an exquisite cheese made by Tim Stone, Galen Stone's great-grandson.

VINAIGRETTE

3 tablespoons white balsamic vinegar

½ cup extra-virgin olive oil

1 small shallot, minced

½ teaspoon minced garlic

⅛ teaspoon salt

⅛ teaspoon sugar

½ teaspoon Dijon mustard

Freshly ground black pepper

SALAD

8 cups mixed greens

2 red Anjou pears, thinly sliced

½ red onion, thinly sliced

½ cup crumbled sharp blue cheese

⅓ cup almonds, toasted

½ cup dried cranberries

1. To make the vinaigrette: Shake together the vinegar, oil, shallot, garlic, salt, sugar, mustard, and pepper in a small jar.

2. In a larger container, combine the greens, pears, onion, cheese, almonds, and cranberries. Toss the salad with the vinaigrette when you're ready to start your picnic.

SERVES 4-6

LITTLENECKS

The first English settlement of Marion was at the head of Sippican Harbor, in an area called Little Neck, because it was so rich in clams. The area reminded the Pilgrims of where they came from in England, with oysters in the salt marshes of the harbor and the necks, and clams, scallops, and quahogs abundant in the bay. To this day, the small hard-shell clams are called "littlenecks."

Boston Salad WITH Blood Oranges AND Red Onion

A member of the butterhead family, Boston lettuce has a soft, buttery leaf that contrasts nicely with its crispy stem. It pairs well with oranges (or other fruit) and red onion.

	Juice of 1 blood orange (¼ cup)
2	teaspoons red wine vinegar
½	cup extra-virgin olive oil
¼	teaspoon salt
⅛	teaspoon black pepper
2	blood oranges
½	cup toasted walnuts, roughly chopped
½	cup sliced red onion
2	small heads Boston lettuce, leaves torn

1. Shake together the orange juice, vinegar, oil, salt, and pepper in a small jar with a tight-fitting lid until blended.

2. Peel and cut the pith off the oranges. Over a container large enough to hold the salad, segment the oranges, catching any juice. Add the walnuts and red onion, and lay the lettuce on top. (You could also transport the lettuce separately in the plastic "clamshell" container they come in at the grocery store.) At your picnic site, combine the salad and dressing when you're ready to eat.

SERVES 4

Coleslaw *with* Carrots *and* Currants

Coleslaw gets a bad rap. Yet made with a deft touch, coleslaw can be crisp, tangy, and savory, with just enough creaminess to pull it all together. This one is a sweet-savory complement to many fishy meals.

3	cups thinly sliced green cabbage
½	cup grated carrots
⅓	cup currants
½	cup mayonnaise
1	tablespoon apple cider vinegar
¼	teaspoon salt
	Freshly ground black pepper
	Pinch of sugar

1. Put the cabbage, carrots, and currants in a large bowl.

2. Whisk together the mayonnaise, vinegar, salt, pepper, and sugar in a small bowl, then pour the dressing over the slaw. Toss to coat.

SERVES 4

THE GRAND YACHTS OF SUMMER

Although ports like Newport are famous for early-20th-century gilded mansions and motor yachts, the quiet harbors are equally rich in maritime history, and many small harbors like Marion, Massachusetts, were home to grand yachts in the days of prewar yachting. Avid yachtsman Galen L. Stone's *Arcadia* was typical: 188 feet long, with a 27-foot beam, five staterooms, and five small boats on deck, including a 33-foot launch. Just down the road, Richard Hoyt owned *Teaser*, a high-speed commuter yacht that in 1925 could beat the *20th Century Limited* train from Manhattan to Albany by 27 minutes. Boats were built along the coast by names that comprise a who's who of yacht builders: C. Raymond Hunt, Nathanael Herreshoff, Olin Stephens, Joel White. Like many of the prewar yachts, the *Arcadia* was pressed into service by the Coast Guard and became a Canadian escort vessel during World War II, then a ferry. She was scrapped in 1969.

Cranberry-Cauliflower-Walnut Salad

I am besotted with Yotam Ottolenghi's *Jerusalem,* a cookbook that explores the flavors of Jerusalem. This recipe, which is loosely adapted from one in the book, makes use of our local cranberries (see page 336). It's a versatile salad — good with fish at dinner, or the next day on a picnic lunch at the beach.

2	cups cauliflower florets
4	tablespoons extra-virgin olive oil
½	teaspoon salt
	Freshly ground black pepper
¼	cup walnuts
2	celery stalks, chopped
¼	cup chopped parsley
¼	cup chopped cilantro
¼	cup dried cranberries
1	tablespoon balsamic vinegar
2	tablespoons extra-virgin olive oil
	Pinch of allspice

1. Preheat the oven to 425°F (220°C).

2. Toss the cauliflower florets with the olive oil, salt, and pepper to taste, and roast on a baking sheet until golden brown, about 30 minutes.

3. Lower the heat to 350°F (180°C) and toast the walnuts until browned and fragrant, about 10 minutes (or use a toaster oven). Set them aside to cool, then coarsely chop with a mezzaluna or knife.

4. Put the roasted cauliflower in a large salad bowl. Add the walnuts, celery, parsley, cilantro, cranberries, vinegar, oil, and allspice, and toss. Serve the salad at room temperature.

SERVES 4

Blueberry-Lemon-Ginger Scones

The glacial fields and meadows of Maine have provided an ideal combination of climate and soil for these antioxidant-rich berries to flourish for thousands of years. Packed with vitamins and intense flavor, native wild low-bush blueberries taste incredibly good, and are great for you.

2¼	cups all-purpose flour
⅓	cup brown sugar
1	teaspoon baking powder
½	teaspoon baking soda
⅓	cup crystallized ginger, minced
1	teaspoon lemon zest
½	teaspoon salt
½	cup (1 stick) cold butter, diced
½	cup fresh blueberries
¾	cup plus 1 tablespoon heavy cream
1	egg
1	teaspoon granulated sugar
½	teaspoon vanilla extract

1. Preheat the oven to 400°F (200°C).

2. In the bowl of a food processor fitted with the metal blade, combine the flour, brown sugar, baking powder, baking soda, ginger, lemon zest, salt, and butter, and pulse to mix.

3. Transfer the mixture to a bowl and use your fingers to gently add the blueberries. Add the ¾ cup cream and stir until the dough comes together, taking care not to overmix. Turn the dough onto a lightly floured work surface and pat into an 8-inch disc. Cut the disc into 8 wedges and place them 2 inches apart on an ungreased baking sheet.

4. In a small bowl, combine the 1 tablespoon heavy cream with the egg, granulated sugar, and vanilla, and beat lightly. Brush the scones with the egg wash. Bake for 20 minutes, or until golden.

MAKES 8 SCONES

OPENING UP THE COTTAGE

Massachusetts Senate president Bill Bulger once teased Governor William Weld about his ancestors having come over on the *Mayflower*. "Actually, they weren't on the *Mayflower*," Weld retorted. "They sent the servants over first to get the cottage ready."

Homemade Lemonade

There's something eminently satisfying about making sure your kids are getting the healthiest ingredients in their drinks. With homemade lemonade, not only is it just three ingredients, but you can regulate the amount of sugar. I'm rather fond of a tart lemonade myself. You can spruce it up with a sprig of mint, thyme, basil, or lavender.

5	lemons
5	cups cold water
¾–1	cup superfine sugar

Roll each lemon on the counter with your hand to release the juices, then cut lemons in half and juice them. Combine the juice with water and sugar in a pitcher. Stir and chill.

SERVES 4-6

VARIATION: CRANBERRY LEMONADE
Enjoy the tart, clean taste of cranberries in your lemonade by stirring ½ cup cranberry juice into 4 cups Homemade Lemonade. Garnish with fresh mint sprigs.

Lavender Lemonade

While salt spray and wind can damage many plants and flowers in a seaside garden, lavender thrives by the sea. My garden has half a dozen lavender plants, which not only give me herbs for food and aromatics, but also a sweet scent; after a hard rain, you can smell the fragrant odor wafting through the sea air.

This is a cooling drink on a hot day — the tartness of the first sip gives way to a surprising floral lavender ending. It goes well with the Curried Lobster Rolls (see page 236).

4	**cups Homemade Lemonade (see page 271)**
4	**tablespoons Lavender Syrup (see page 322)**

Combine the lemonade with the lavender syrup in a pitcher and serve over ice.

VARIATION: SPARKLING LEMONADE

Make this lemonade into a refreshing spritzer by adding a liter of seltzer water.

Fresh Limeade

This refreshing limeade is delicious at lunch (it'd be great with a lobster roll), or as an afternoon pick-me-up. Muddle some mint sprigs in the glass and add a splash of rum and you're on your way to a new kind of mojito, too. This recipe makes a single serving, but you'll have a big batch of simple syrup that you can use to sweeten many drinks — lemonade, herbal tea, iced coffee, for example.

1½ limes
3   cups water
1   cup sugar
    Club soda
    Sprig fresh thyme or mint

1. Squeeze the juice from the limes.

2. Combine the water and sugar in a saucepan and bring to a boil, stirring to dissolve the sugar. Let cool. (You'll put most of the simple syrup in a jar in the refrigerator for future use.)

3. Fill a short rocks glass with ice and add the lime juice, then equal parts club soda and simple syrup. Serve garnished with a sprig of thyme.

SERVES 1

Thyme-Lemon Limeade

A glass of limeade infused with thyme and lemon turns it from a lemonade-stand drink to a libation you could serve in the evening spiked with alcohol (or not).

12 fresh thyme sprigs
½   cup lime juice
½   cup lemon juice
1   cup superfine sugar
5   cups water

Using a wooden spoon, muddle 8 of the thyme sprigs in the bottom of a pitcher. Add the lime juice, lemon juice, sugar, and water, and stir until the sugar is dissolved. Serve in glasses over ice, garnished with the remaining 4 thyme sprigs.

SERVES 4

Lemon Ginger Tea

This revitalizing tea is good for what ails you — be it a cold, seasickness, or a case of the chills. Served with a bit of chocolate, this tea also concludes a satisfying dinner party on a bright, refreshing high note.

8   cups water
1   (4- to 6-inch) piece ginger root, peeled and grated
4   lemons, squeezed, plus 1 lemon, sliced, for garnish (optional)
1   cup or more ginger syrup (recipe follows)

1. Combine the water, ginger root, and lemon juice to a boil in a pot. Remove from the heat and steep for 15 to 20 minutes.

2. Strain and stir in the simple syrup. Taste and add more sweetener if you like. Serve hot or cold. The tea may be kept covered in the refrigerator for up to 1 week.

SERVES 8

Ginger Syrup

Simple syrup is a sweetener that adds volumes to a drink. Infused with ginger, this simple syrup is not only healthy but lifts many drinks, from cocktails to tea, to a new level. A glass of seltzer on ice would also be bolstered by a tablespoon of ginger syrup.

1   cup water
1   cup sugar
6   (¼-inch-thick) rounds of ginger, peeled

Combine the water, sugar, and ginger in a saucepan and bring to boil, then simmer 10 minutes. Cool and strain into a glass jar. This simple syrup will keep for 1 month refrigerated.

MAKES 1⅓ CUPS

"I'm fond of anything that comes out of the sea — and that includes sailors."

— JANET FLANNER

Del's Lemonade

What is it about regional flavors from childhood that brings back vivid memories? To Rhode Islanders, it may be a frozen slushy lemonade from Del's. Rhapsodized one woman: "Del's is my childhood, the perfect thing on a hot summer day, my happy place.... You know it's summer once you've had your first Del's of the season... anytime I go back to Rhode Island in the summer, the first thing I need to do is have about 20 Dels." Similar to a granita, Del's lemonade can be found at food trucks by the beach, along crowded beach highways, anywhere there's heat and summer. It's sunshine in a glass.

¾   **cup sugar**
4   **cups water**
     Juice of 4 lemons with pulp (about 1 cup)
1   **teaspoon lemon zest**

1. Put a 9- by 12-inch glass or porcelain baking dish in the freezer for 30 minutes.

2. In a medium bowl, dissolve the sugar in the water, mixing thoroughly, add lemon juice and zest, and stir. Taste and add more sugar or lemon if desired.

3. Pour the lemon mixture into the chilled dish and freeze for 90 minutes, stirring briefly every 20 minutes or so.

4. Scoop into the bowl of a food processor and purée until smooth, 20 to 30 seconds. Serve right away!

SERVES 4

HOBIE

Known as the Henry Ford of surfboarding, Hobie Altier designed the Hobie Cat, the ultralight fiberglass catamaran that put sailing on another plane.

I first sailed one in 1969. I was just a kid, but my older cousin Lynn (who drove a powder-blue Mustang and was in college) let me sail her Hobie 14. It took off like lightning on the WeWeantic River, skimming the surface like a firefly. We all cracked up on the shore when she later took it out with her black Lab mutt and the webbing came loose on the trampoline-style deck. She and her startled dog fell through the canvas while the boat kept on sailing.

More recently, I sailed a Hobie Mirage Adventure Island: part kayak, part trimaran sailboat, part paddleboat. Slung low in the molded cockpit, inches from the waves, you feel free and fast as you skim across the water, your leeward ama chiseling into the water while the windward one goes airborne. You surf, you sail, you sing (well, I sing), you soar — you just can't believe your good fortune to be out on the water, so close to the water, so influenced by the wind, on such a beautiful day. It's a blast.

White Chocolate–Macadamia Nut Cookies

While my kids would argue that macadamia nuts are an acquired taste, most adults will find these cookies grand on a boat or beach picnic.

1 cup (2 sticks) butter, softened	½ teaspoon salt
½ cup firmly packed brown sugar	1 teaspoon baking soda
½ cup granulated sugar	2½ cups all-purpose flour
2 large eggs	1 cup white chocolate chips
1 teaspoon vanilla extract	1 cup macadamia nuts, coarsely chopped

1. Preheat the oven to 375°F (190°C).

2. Cream together the butter, brown sugar, and granulated sugar in a large bowl. Add the eggs and vanilla, beating until smooth. Add the salt, baking soda, and flour, and blend until just mixed. Stir in the white chocolate chips and nuts.

3. Using a teaspoon, drop the batter onto baking sheets lined with parchment paper, spacing them 1 inch apart. Bake for 10 minutes, and then transfer the cookies on the parchment paper to a cooling rack. After 5 minutes, remove the parchment and let the cookies cool completely on the rack.

MAKES 3½ DOZEN

Green Tea Granita with Ginger Syrup and Mint

This tangy, refreshing granita is easy to make and features ginger as an added benefit. Since ancient times, ginger has been used as a remedy against digestive problems. (Indeed, ginger ale evolved from a ginger beer made in colonial America as an antidote to nausea.)

3	cups water
4	green tea bags
½	cup local honey
½	cup sugar
1	teaspoon lemon juice
½	teaspoon grated ginger
	Mint sprigs, for garnish

1. Bring the water to a boil and steep the tea bags for 15 to 20 minutes. Remove the bags, add the honey, sugar, lemon juice and ginger, stir, and allow to cool.

2. When cool, pour the liquid into an 8- by 8-inch glass or plastic dish, and freeze for 2 hours. After 2 hours, fork it to break up the ice and any lumps. Every 30 minutes, fork it so it turns into a smooth, icy sorbet, about 4 hours. (If it gets too hard, set on counter for 15 minutes before serving.)

3. Garnish with mint and serve.

SERVES 4

"WAKING OR SLEEPING, I DREAM OF BOATS."

— E. B. WHITE

Honeysuckle Sorbet

It sounds corny to say it, but along the gentle coves, harbors, and inlets of New England in the summertime, the scent of beach roses and honeysuckle wafts through the air on afternoon breezes. This sorbet tastes like what you would imagine the smell of honeysuckle to taste like. It benefits from a swirl of Campari. If you plan to make it and serve it the next day, be forewarned that it gets sweeter and less nuanced — it's best the day you make it. The blossoms need to be soaked overnight to make the syrup, but the sorbet itself is best served the day it's made. Kept overnight the flavor becomes sweeter and less nuanced.

1	cup honeysuckle blossoms, lightly packed
2	cups cool water plus 1 cup water
1	cup sugar
½	teaspoon lemon juice
½	teaspoon lime juice
	Dash of cinnamon
	Dash of nutmeg
	Honeysuckle flowers, for garnish

1. Remove the green bottom stem from the honeysuckle, and soak the blossoms in the 2 cups cool water in a glass container overnight, covered lightly on the counter.

2. The next day, drain and reserve the liquid.

3. Make a simple syrup by heating the 1 cup water with the sugar in a saucepan, stirring, until the sugar is completely dissolved, about 5 minutes. Remove from the heat and add the lemon and lime juice, then set the simple syrup aside to cool.

4. Combine the honeysuckle water with the cooled simple syrup, add the cinnamon and nutmeg, and pour into a 13- by 9-inch pan. Freeze for 2 hours and then rake the surface with a fork to break up the ice and any lumps.

5. Continue to freeze and fork the sorbet every 30 minutes until it turns into a smooth, icy sorbet, about 4 hours. (If it gets too hard, set it on the counter for 15 minutes before serving.) Garnish with honeysuckle flowers before serving.

SERVES 8

NOTHING UNNECESSARY

If you've sailed Maine's waters, or simply enjoyed the pleasures of sitting by the sea, you'll understand why certain folk embrace a lifelong love of Maine and don't give a whit about going south of the border ("I get a little jittery when I go past Kittery," said *Bert and I*). Bracing wind, bone-numbing water, crystallized views that look like a photo filtered through Instagram — Maine is unique, from clamming on Casco Bay, to popovers at the Jordan Pond House, to hiking Bubble Rock. It gets in your blood.

Where else could you find a town like Brooklin (population 824), which has nine boatbuilding operations? (The graduation requirement from the local elementary school involves eighth graders going into the forest, claiming a log, taking it to the sawmill and then to the boatshop to make a boat.) One of those boatyards was started by naval architect Joel White, son of E. B. White, who created simple boats (including the Nutshell Pram and the Haven 12½) for four decades until his death in 1997. "In a world gone crazy for gadgets and goo-gaws on boats," opined *Wooden Boat* magazine upon his death, "he preserved a sense of elegance and purity. He was not interested in self-aggrandizement. His boats were adorned with nothing unnecessary."

ON THE
SIDE

$\mathbf{T}$HIS CHAPTER is about side dishes that go well with seafood, although I've been known to turn two or three side dishes into a meal. These vegetable-inspired dishes are wonderful to have at your disposal after a particularly rewarding farmers' market visit.

These days, many towns and communities take advantage of fresh ingredients that reflect the terroir of a region. Take Westport, Massachusetts, for example. At first glance, Westport looks like much of southeastern Massachusetts: rolling fields, cows grazing, old stone walls, tangles of rose hips, and ospreys (the buzzards of Buzzards Bay) circling overhead. But there's something unique about Westport and its environs, for the two branches of the Westport River and the sea air

create the longest growing season in all of New England. The loam is fantastic. Like in many communities, early settlers followed the lead of Native Americans, raising corn and beans. There are native cranberries. Westport Point is tiny, but has a functioning fishing village, with fishing boats and draggers. The oysters are renowned. An independent spirit prevails among small farms, several of which are organic.

Years ago, Westport was the chop-shop capital of the world, but in 2000, a motion was passed at a town meeting to make the town into the farming community of the south coast. Today, Westport is a designated right-to-farm community. Aggie artisans keep cows, make cheese, raise turkeys, stomp grapes, and grow organic vegetables, servicing chefs as far away as Boston and New York, not to mention regional chefs and home cooks who love preparing meals with the local bounty.

This chapter includes dishes that go well with seafood as sides and can stand on their own as appetizers or lunch. They celebrate the flavors of coastal New England.

Summer Arugula Salad *with* Roasted Corn Salsa *and* Cilantro Vinaigrette

You're going to love the crispy, salty baked-prosciutto crumble in this salad. You may want to make extra and save for other dishes, as it tastes great on so many foods!

ROASTED CORN SALSA

3	ears husked corn
2	plum tomatoes
½	jalapeño, seeded
¼	cup diced red onion
1	tablespoon chopped cilantro
1	tablespoon lime juice
1	teaspoon cumin
¼	teaspoon salt
	Freshly ground black pepper
3	slices prosciutto

CILANTRO VINAIGRETTE

6	tablespoons extra-virgin olive oil
2	tablespoons lemon juice
½	teaspoon white wine vinegar
2	teaspoons chopped cilantro
¼	teaspoon salt
⅛	teaspoon black pepper
4	cups baby arugula
1	ripe Hass avocado, sliced

1. Preheat the broiler.

2. To make the salsa: Place the corn, tomatoes, and jalapeño on a baking sheet and broil for 1 minute. Turn the tomatoes and jalapeño over and broil for an additional minute. Remove the tomatoes and jalapeño and set aside. Continue to broil the corn, turning frequently, until lightly charred all over, 8 to 10 minutes. When cool enough to handle, cut the kernels from the cobs. Chop the tomatoes and jalapeños, then combine with the onion, cilantro, lime juice, cumin, salt, and pepper to taste in a small bowl.

3. Turn the oven down to 400°F (200°C). Lay the prosciutto on a lightly oiled baking sheet and bake for 5 minutes, watching that it doesn't burn. When cool enough to handle, crumble the meat into small morsels.

4. To make the vinaigrette: Whisk together the oil, lemon juice, vinegar, cilantro, salt, and pepper.

5. Put the arugula in a salad bowl and toss with the dressing; start with about three-quarters of the dressing to taste, adding more if desired. Assemble on individual plates, and top each salad with some salsa, a fanning of avocado, and a sprinkling of prosciutto.

SERVES 4

Pickled Daylily Buds

My friend Jody can pickle, can, and cook just about anything, and never lets anything go to waste. She's the kind of gal who buys a big cut of meat and butchers it herself, and makes her own horseradish and curries. Many people have orange daylilies growing in their garden or wild in the woods, but only Jody would look at this common flower and think: here's an opportunity.

Serve these buds as an appetizer or with sandwiches. They also are delicious as an accompaniment to strong cheddar cheese. They will keep for about 2 months refrigerated.

2½	cups water
4	tablespoons salt
35	daylily buds (the tastiest are those that are just about to open)
1	cup apple cider vinegar
1	clove garlic, halved

1. Place 2 cups of the water and the salt in a bowl, stirring until the salt dissolves. Add the daylily buds and let stand overnight, covered.

2. In a small saucepan, heat the vinegar and garlic.

3. Put the drained daylily buds in a clean pint jar. Add the hot cider to almost the top, cover, and allow to cool on the counter. When cool, place in refrigerator and leave for 2 weeks to pickle.

MAKES 35 PICKLES

SAUTÉED DAYLILY BUDS

Wash and dry a cup or two of daylily buds. Sauté a bit of butter and olive oil until they are slightly browned, adding a minced shallot if you have it. Then add the buds and sauté until softened. Season with salt and pepper to taste.

Lemon-Kale Salad

People feel about kale much like they do about politics: opinionated and divided. If I mention that kale's on the menu for dinner, my kids groan or roll their eyes and look at me like I'm a weirdo. My secret is: chop it up fine, and it becomes a stealth missile of supernutrition and an eye-popping green wonder without having to bear the health-food-nut rap. This much-loved, often-requested salad is on the menu at Allium in Great Barrington, Massachusetts. Serve with salmon for the ultimate power couple combo.

CROUTONS

½	loaf sourdough bread
¼	cup olive oil
	Salt

DRESSING

1	garlic clove, peeled
1	tablespoon Dijon mustard
⅓	cup lemon juice
½	cup grapeseed oil
½	cup extra-virgin olive oil
1	cup grated Pecorino Romano cheese
1	Granny Smith apple
	Juice of ½ lemon
1½	pounds curly kale

1. Preheat the oven to 350°F (180°C).

2. To make the croutons: Cut the bread into tiny cubes, place in a medium bowl, and toss with the olive oil and salt. Place on a baking sheet and bake for 5 to 7 minutes, until golden. Turn, and bake another 5 to 7 minutes.

3. To make the dressing: In a food processor, crush the clove of garlic, then add the mustard and lemon juice. With the processor on, slowly drizzle in the grapeseed and olive oils until emulsified. Add ½ cup of the cheese and pulse until blended.

4. Core the apple and dice into tiny cubes. To prevent browning, toss the apple cubes with lemon juice.

5. To prepare the kale, strip out the stem of each kale leaf; wash the leaves and spin in a salad spinner to remove all excess water. Chiffonade (see page 53) or cut the leaves into thin strips. Put the kale in a large bowl with the apples and croutons. Shake up the dressing, then pour over the salad. Add the remaining ½ cup cheese and toss vigorously. Serve immediately.

SERVES 6

Sautéed Cauliflower with Scallions and Currants

Cauliflower holds its own among strong flavors. It's more versatile than broccoli and complements assertively flavored fish (like salmon) well.

1	medium head cauliflower, divided into florets
½	teaspoon cumin seeds
1	tablespoon extra-virgin olive oil
1	tablespoon chopped garlic
	Salt and freshly ground black pepper
⅓	cup dried black currants, soaked in water for 10 minutes and drained
1	tablespoon minced scallion greens
	Juice of ½ small lemon

1. Fill a pot with a couple inches of water and bring to a boil. Put the cauliflower in a steaming basket and steam until just tender, 10 to 12 minutes.

2. Toast the cumin seeds in a dry cast-iron skillet until aromatic, about 1 minute. Set aside. Heat the olive oil in the same skillet over medium heat. Add the garlic and sauté until fragrant, about 1 minute. Season with salt and pepper, and then add the cauliflower and raise the heat to medium-high. Sauté, stirring occasionally, until the cauliflower is golden brown, about 10 minutes.

3. Turn the cauliflower into a serving bowl and stir in the currants, scallions, cumin seeds, and lemon juice. Season with salt and pepper to taste.

SERVES 4-6

Dandelion Greens, Oranges, AND Goat Cheese

Appearing in early spring, dandelion greens are one of the most nutritious leafy greens. For some, picking and eating dandelion greens is a beloved spring ritual. This dish goes well with salmon, sardines, and other oily fish.

DRESSING

4	tablespoons extra-virgin olive oil
2	tablespoons grapeseed oil
2	tablespoons red wine vinegar
1	tablespoon lime juice
¼	teaspoon salt
	Freshly ground black pepper

SALAD

8	cups chopped (1-inch length) dandelion greens
½	small red onion, thinly sliced
½	cup toasted pecans
½	cup cranberries
2	seedless navel or blood oranges, segmented
3	ounces goat cheese, crumbled

1. To make the dressing: Whisk together the olive oil, grapeseed oil, vinegar, lime juice, salt, and pepper to taste in a small bowl.

2. To make the salad: Put the dandelion greens, red onion, pecans, and cranberries in a serving bowl and toss with the dressing. Add the orange segments, sprinkle with goat cheese, and serve.

SERVES 6-8

Arugula  Bartlett Beet Salad

The oldest farm on Nantucket, Bartlett's Farm has more than 125 acres that have been tilled for seven generations on what the Algonquin called "the faraway land."

4	small Bartlett's or other golden beets	⅓	cup extra-virgin olive oil
1	bunch arugula	2	tablespoons lemon juice
1	small red onion, thinly sliced	½	teaspoon salt
1	ripe Hass avocado, sliced		Freshly ground black pepper
¼	cup toasted pine nuts	10	thin slivers Parmigiano Reggiano or Parmesan

1. Preheat the oven to 425°F (220°C).

2. Wrap the beets in foil and roast until soft when pierced with a knife, about 1 hour. When cool, cut off the ends and slip off the skins. Slice, and toss with the arugula, onion, avocado, and pine nuts in a bowl.

3. Whisk together the oil, lemon juice, salt, and a generous amount of pepper in a small bowl. Pour the dressing over the salad, toss lightly, and garnish with slivers of cheese.

SERVES 4

Golden Beet Salad *with* Gorgonzola

Bright, earthy golden beets pair beautifully with Gorgonzola cheese. This is a pretty salad that's delicious with grilled shrimp. You could serve the beets on a bed of arugula or lettuce to stretch the salad.

3	medium golden beets
1	tablespoon finely minced red onion
2	tablespoons white wine vinegar
6	tablespoons extra-virgin olive oil
1	tablespoon Dijon mustard
½	teaspoon sea salt
¼	teaspoon freshly ground black pepper
½	teaspoon minced fresh thyme
¼	cup roasted, salted pistachio nuts, shelled and coarsely chopped
2	ounces Gorgonzola cheese

1. Preheat the oven to 425°F (220°C).

2. Wrap the beets in foil and roast until soft when pierced with a knife, about 1 hour. When cool, cut off the ends and slip off the skins.

3. In a medium bowl, whisk together the onion, vinegar, oil, mustard, salt, pepper, and thyme, then toss with the beets. Marinate in the refrigerator for 1 to 2 hours.

4. When ready to serve, add the nuts and cheese and toss.

SERVES 4

Sesame Spinach

Much touted for its nutritional heft, spinach is rich in iron, which helps
carry oxygen to your muscles and brain, restoring energy and increasing your vitality.
It's also a great source of folic acid, and vitamins K, A, C, and B_2. Serve spinach
with foods rich in vitamin C (roasted tomatoes, bell peppers, broccoli, orange slices)
for increased iron absorption.

½	tablespoon sesame seeds
1	tablespoon sunflower seeds
2	teaspoons olive oil
1	garlic clove, thinly sliced
2	scallions, thinly sliced
⅛	teaspoon red pepper flakes
8	ounces baby spinach
¼	teaspoon salt
	Freshly ground black pepper

1. Heat a dry skillet over medium heat. Add the sesame and sunflower seeds
and toast until they are fragrant and golden. Remove the seeds.

2. Add the olive oil to the skillet, and sauté the garlic and scallions for
1 minute. Add the pepper flakes, spinach, salt, and pepper to taste;
toss to combine and cook until wilted, about 3 minutes. Toss with the
seeds and serve.

SERVES 2

Roasted Tomatoes

This versatile dish goes well with oven-roasted or grilled fish. Roasted tomatoes form a good base for couscous, tomato sauce . . . all sorts of things. Plus, it's so easy. If you've never made these, you have no idea. Make a big batch: they freeze well and you'll find yourself turning to them often. Wait no longer.

12	medium tomatoes, halved
½	cup extra-virgin olive oil plus 1 tablespoon
1½	teaspoons kosher salt
2	tablespoons minced garlic
2	teaspoons fresh thyme
	Freshly ground black pepper

1. Preheat the oven to 450°F (230°C). Line a rimmed baking sheet with foil and brush with the 1 tablespoon olive oil.

2. Place the tomatoes skin side down on the prepared baking sheet. Drizzle the ½ cup olive oil over the tomatoes and sprinkle the salt, garlic, thyme, and pepper over the top. Bake for 1 hour or until the tomatoes are roasted and the garlic is golden. Serve immediately or at room temperature.

SERVES 6

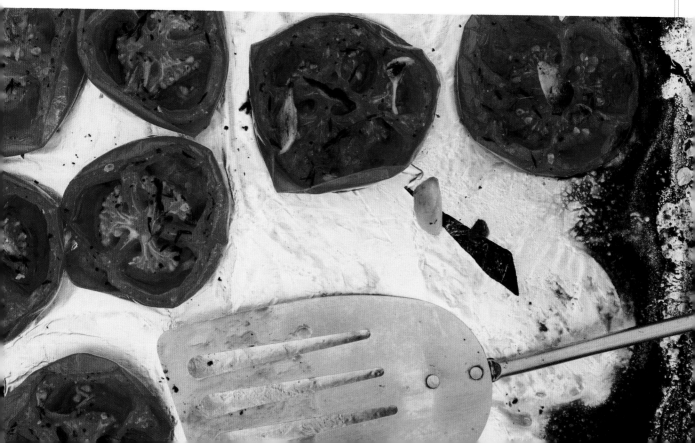

Pomegranate–Wild Rice Salad WITH Roasted Squash AND Greens

This recipe is a riff on one I saw in *Bon Appétit* magazine. I serve it at Thanksgiving, loving the color and the counterpoint of the textures, but it's good year-round, served on its own or with any white fish.

1	pomegranate (½ cup seeds)
1	cup wild rice
	Kosher salt
2	cups butternut squash, peeled, seeded, and cut into bite-size pieces
5	tablespoons olive oil
½	teaspoon salt
	Freshly ground black pepper
2	tablespoons champagne vinegar
2	scallions, thinly sliced
2	tablespoons chopped parsley
½	cup toasted sliced almonds
1	cup loosely packed baby arugula
½	cup finely chopped kale, loosely packed
¼	cup crumbled feta cheese

1. Cut the pomegranate in half. Place one half cup-side down on your palm and tap with a wooden spoon to remove the seeds. Repeat with the other half.

2. Preheat the oven to 425°F (220°C).

3. Bring a large pot of salted water to a boil, add the rice, and cook until tender, about 25 minutes; drain and rinse. Spread the rice out on a baking sheet to dry and cool.

4. Toss the squash with 2 tablespoons of the oil, salt, and pepper to taste, and lay the pieces out on a rimmed baking sheet. Roast for 15 to 20 minutes, turning once, until golden brown and tender. Cool.

5. In a salad bowl, whisk together the remaining 3 tablespoons oil, the vinegar, salt, and pepper; then add the wild rice, squash, scallions, parsley, pomegranate seeds, and almonds, and toss. Add the arugula and kale, toss again, sprinkle with feta, and serve.

SERVES 4-6

Roasted Cauliflower *with* Agave *and* Smoked Paprika

A beautiful truth on a Tuesday night in winter when your kids are hungry: this dish can be assembled in less than 15 minutes. Then, while the cauliflower roasts for 30 minutes, you can do the hundred other things you need to tend to before dinner. This is a satisfying accompaniment to grilled fish and couscous.

1	head cauliflower, separated into small florets
3	tablespoons olive oil
1	shallot, minced
1	teaspoon sea salt
	Freshly ground black pepper
2	tablespoons agave nectar or honey
¼	teaspoon smoked paprika
	Pinch of red pepper flakes
½	teaspoon lemon zest

1. Preheat the oven to 450°F (230°C).

2. In a bowl, gently toss the cauliflower with the oil, shallot, salt and pepper, agave nectar, paprika, pepper flakes, and lemon zest. Spread the coated cauliflower on a baking sheet and roast uncovered for 30 minutes, or until tender.

SERVES 4-6

"Out on the islands that poke their rocky shores above the waters of Penobscot Bay, you can watch the time of the world go by, from minute to minute, hour to hour, from day to day, season to season."

— ROBERT MCCLOSKEY, *TIME OF WONDER*

Swamp Yankees

To the outside world, both my parents' families were cut of the same cloth — their Protestant forebears had arrived in the 1600s and shared a resourcefulness, sense of humor, and thrift that helped generations survive and occasionally thrive in New England.

That's where the similarities ended.

My father's ancestors were captains of sea and industry in the 19th and early 20th centuries. While I have no memory of anyone in the family ever uttering the word "WASP," let alone "Brahmin," there were code words in conversation that indicated one's background or status. Crow Point. Fly Club. Andover. *The* Country Club. Simple words, loaded with meaning, possibility, and privilege. As my grandmother (a dead ringer for Maggie Smith in *Downton Abbey*) remarked when I sat for my Bachrach engagement photo, *The New York Times* includes your grandparents' names in the engagement announcement merely to indicate where the money came from.

My mother hailed from stock that turned out the teachers, tanners, ministers, and volunteer soldiers who populated New England. While her ninth great-grandfather Nicholas Easton governed the Rhode Island Colony in 1672 (it wouldn't become a state for another hundred years), it was only after he'd been disarmed and banished as a rogue minister from Massachusetts Bay.

Easton arrived from Plymouth, England, in 1634 aboard the *Mary and John*, one of 250 "Pilgrim Ships" that transported 7,100 intrepid travelers from England in the early 1600s. Their descendants are some of the Yankees who still live modestly today in boggy Southeastern Massachusetts a few miles from Plymouth Rock. Swamp Yankees.

Charred Broccolini with Lemon and Red Pepper Flakes

You can prepare this dish on the grill or in the oven (on a rimmed baking sheet under the broiler). Or pan-fry broccolini, stirring the anchovies and garlic into the olive oil for a classic Italian combination.

1	bunch broccolini, ends trimmed and leaves removed
1	lemon, halved
1	clove garlic, minced
2	teaspoons anchovy paste
1	tablespoon extra-virgin olive oil
	Pinch of red pepper flakes
¼	teaspoon coarse salt
	Freshly ground black pepper
¼	cup shaved Parmesan cheese

1. Prepare a medium-hot fire in a gas or charcoal grill.

2. Wash and dry the broccolini. Toss the broccolini with the juice of half the lemon, the garlic, anchovy paste, oil, pepper flakes, salt, and pepper to taste.

3. Lay the broccolini flat on a grill pan and grill for 4 minutes per side, turning gently halfway through. When you turn it, lay the remaining half lemon cut-side down on the grill to char that as well. Place on a serving platter and sprinkle the Parmesan over the top.

4. Serve immediately with the charred lemon, which can be used to squeeze more juice on the broccolini.

SERVES 2

Pico de Gallo

Pico de gallo (which means "rooster's beak" in Spanish) is a fresh, uncooked Mexican salad traditionally made with chopped tomatoes, serranos, and lime. Here, use jicama and cucumbers, which, when tossed with sweet melon, pungent mint, and the bite of fresh chiles, creates an explosion of flavor and a fresh endnote to a big meal. I like my salad after the main course — this one works particularly well after seafood paella. It also makes a refreshing summer appetizer.

1	small red onion, thinly sliced
4	tablespoons fresh lime juice
1	teaspoon salt
1	garlic clove, minced
3	small Kirby cucumbers, peeled, halved, and thinly sliced
½	jicama, peeled and thinly sliced into matchsticks
2	seedless oranges, peeled, segmented, and cut into quarters
1	jalapeño, halved, seeded, and thinly sliced crosswise
2	tablespoons diced red bell pepper
1	tablespoon chopped fresh mint
1	bunch arugula, large stems removed
1	firm, ripe green melon or cantaloupe, peeled, seeded, and cut into thin crescents

1. Put the onion in a heat-proof bowl and cover with boiling water, letting it stand 2 minutes. Drain and combine with the lime juice, salt, and garlic.

2. Combine the cucumbers, jicama, oranges, jalapeño, and bell pepper. Add the onion mixture and the mint and toss gently. Cover and refrigerate until ready to serve.

3. Garnish each salad plate with a handful of arugula leaves, and then place two cantaloupe slices in an X over the arugula. Spoon the salad on top and serve.

SERVES 6

Grilled Vegetable Salad

This dish is ideal for high-summer garden abundance, and goes well with just about any grilled fish, in particular salmon. You can easily substitute other vegetables from the garden.

1	medium Italian eggplant, peeled and cut into ¼-inch slices
1	medium white or yellow onion, cut into ½-inch rounds
1	red pepper, seeded and cut into quarters
2	medium zucchini, cut into ¼-inch slices
2	yellow squash, cut into ¼-inch slices
2	tablespoons grapeseed oil

DRESSING

¼	cup olive oil
2	tablespoons balsamic vinegar
1	garlic clove, finely minced
2	tablespoons basil chiffonade (see page 53)
1	tablespoon chopped chives
1	teaspoon salt
½	teaspoon black pepper

1. Prepare a medium-hot fire in a gas or charcoal grill.

2. Brush the eggplant, onion, bell pepper, zucchini, and yellow squash pieces with the grapeseed oil.

3. Grill the vegetables until cooked through and lightly charred; 5 to 8 minutes per side for the zucchini and squash, and about 10 minutes per side for the eggplant, onion, and bell pepper.

4. Cut all the vegetables into bite-size pieces, and place in a large bowl.

5. To make the dressing: Put the olive oil, vinegar, garlic, basil, chives, salt, and pepper in a jar with a tight-fitting lid, and shake to mix. Pour the dressing over the vegetables and toss to coat. Let sit for 1 hour at room temperature, then serve.

SERVES 6

Fall Salad *with* Arugula *and* Roasted Root Vegetables

Thirty-five miles south of Boston, the town of Duxbury was inhabited by Native Americans as early as 12,000 B.C., and archaeologists have found evidence of fish harvesting along the shore and wetlands. The Wampanoags (which means "People of the First Light") settled in this lowlying region beginning around 1500, and when European settlers arrived, Wampanoags were already calling it Mattakeesett, or "place of many fish."

There's a captain's house in Duxbury that sits on a thumb of land overlooking Duxbury and Kingston Bays and beyond them into Massachusetts Bay and the Atlantic Ocean. Native Americans once summered where the house stands, fishing the waters and digging clams. The house was once owned by the granddaughter of Joshua Lawrence Chamberlain (a Bowdoin professor who became a Civil War hero after defending Little Round Top at Gettysburg, and later Governor of Maine), and is now a retreat center administered by the local congregational church. Called the Cedar Hill Retreat Center, it offers an intimate retreat space for families and groups in an exquisite natural setting. This recipe is from my cousin Donna, who arrives at family parties at the captain's house with the best food.

ROASTED VEGETABLES

2	medium carrots, diced small
2	medium beets, diced small
1	parsnip, diced small
1	tablespoon extra-virgin olive oil
¼	teaspoon salt
¼	teaspoon dried thyme

SALAD

4	cups loosely packed baby arugula
2	tablespoons shaved Parmesan
¼	cup slivered almonds, toasted
⅓	cup thinly sliced raw zucchini
½	small red onion, thinly sliced
½	cup Roasted Chickpeas (recipe follows)

DRESSING

2	garlic cloves, peeled
	Salt
1	teaspoon Dijon mustard
2	tablespoons white wine vinegar
6	tablespoons extra-virgin olive oil
	Freshly ground black pepper
1	tablespoon finely chopped shallot

continued on next page

1. Preheat the oven to 400°F (200°C).

2. To roast the vegetables: Toss the carrots, beets, and parsnip with olive oil, salt, and thyme, and roast on a baking sheet for 40 minutes.

3. To make the dressing: Boil the garlic cloves in water for 10 minutes, and then drain and cool. In a small bowl, mash the garlic with the salt, mustard, white wine vinegar, olive oil, pepper, and shallot.

4. To assemble the salad: Put the roasted vegetables, arugula, Parmesan, almonds, zucchini, onion, and chickpeas in a salad bowl. Toss gently with the dressing, pouring on as much as you desire (the recipe makes ½ cup). Season with salt and pepper.

SERVES 4

Roasted Chickpeas

Toss these crunchy bites into salads, or add a bit of olive oil and vinegar, and you have a side on its own. Chickpeas are also delicious tossed with roasted red peppers and minced red onions.

1	(15.5-ounce) can chickpeas, drained
1	tablespoon extra-virgin olive oil
½	teaspoon fresh sage chiffonade (see page 53)
½	minced fresh garlic
½	teaspoon dried thyme
⅛	teaspoon sea salt
⅛	teaspoon red pepper flakes

1. Preheat the oven to 400°F (200°C).

2. Combine the chickpeas, olive oil, sage, garlic, thyme, salt, and pepper flakes on a baking sheet. Toss to coat the chickpeas. Bake for 15 minutes, until crisp.

MAKES 1 CUP

Watermelon, Arugula, AND Feta Salad

The secret to this salad is the bounteous flavors that pull you in different directions — salty feta, juicy sweet watermelon, sharp onion, and slightly bitter arugula. Each does its part to create a symphony of flavor. Other options include adding some roasted pine nuts, or substituting blue cheese for the feta. Toss the arugula with the dressing at the last minute to keep it crisp.

4	cups chopped seedless watermelon, chilled
20	grape tomatoes, halved
10	Kalamata olives, pitted and chopped
½	small red onion, thinly sliced and quartered

DRESSING

⅓	cup lime juice
1	tablespoon honey
⅓	cup extra-virgin olive oil
½	teaspoon salt
	Freshly ground black pepper
3	cups baby arugula
½	cup whole mint leaves, julienned
¾	cup crumbled feta

1. Assemble the watermelon, tomatoes, olives, and red onions in a large bowl.

2. To make the dressing: Whisk together the lime juice, honey, olive oil, salt, and pepper to taste in a small bowl.

3. Add arugula, mint, and feta to the salad and toss with the dressing. Serve immediately.

SERVES 6

Seaweed Salad

The cold, clean waters of midcoast and Downeast Maine yield amazing seafood that long ago gained cult status, particularly lobsters, oysters, and scallops, but there are other foods that come from the sea — such as seaweed, which is harvested in the wild.

1	ounce dried wakame seaweed
1	tablespoon rice wine vinegar
½	teaspoon toasted sesame oil
1	scallion, minced
½	teaspoon red pepper flakes
¼	teaspoon salt
¾	teaspoon sugar
1	tablespoon sesame seeds, toasted

1. Soak the wakame for 30 minutes in lukewarm water in a large bowl. Drain, squeeze dry, and cut into ½-inch wide strips, if not already sliced.

2. In a small bowl, whisk together the vinegar, oil, scallion, pepper flakes, salt, and sugar. Toss the seaweed with the dressing, sprinkle with sesame seeds, and let stand for 20 minutes or so before serving.

SERVES 4-6

"THE CURE FOR ANYTHING IS SALT WATER — SWEAT, TEARS, OR THE SEA."
— ISAK DINESON

GREEN ACRES

While biking and rambling along the shore and islands Downeast, I happened upon a small roadside hut with a sign advertising bags of seaweed for sale. Behind the lean-to were big sheets of dulse and nori drying on racks in the summer sun, benefited by an ocean breeze. With its rocky coastline, nutrient-rich waters, climate, and dramatic tides, Downeast Maine is an ideal environment for more than 250 species of seaweeds that grow in bands running roughly parallel along the coast and islands in the Gulf of Maine/Bay of Fundy region. Rockweed is one of the state's most valuable marine resources, a $20-million-a-year market.

A delicacy since the time of Confucius, seaweed has been harvested for fertilizer, food, and medicine. Seaweed was so important in ancient Iceland that the laws dictated one's rights to collect seaweed on a neighbor's land. In northern New England, rockweeds have been collected for fertilizer since the 1600s, and more recently, for use in animal food supplements, and for packing and shipping live lobsters.

Demand for sea vegetables is increasing. They are high in calcium, protein, minerals, and iodine (which offers thyroid support after exposure to radiation). As this nutritional powerhouse becomes better known, seaweeds are increasingly being harvested for food. Some Maine sea vegetable farmers harvest seaweed from aquaculture nets strung across the water's surface, rather than collecting the plant wild and venturing out on slippery rocks at low tide in a storm.

Others in this extremely independent state are committed to sustainably harvesting wild seaweed in these cold waters, and go out in wooden boats before dawn when there's an early-morning low tide to the outer islands of the Schoodic Peninsula. Edible seaweed thrives on ledges that break just above the water line on new- and full-moon tides.

The harvesters at Ironbound Island Seaweed, for example, scramble onto exposed rocks, waist-high in the waves, taking care to leave some of the slow-growing plants for regeneration, then return home and hang their seaweed — be it nori, dulse, kombu, alaria — in the sun, where it dries in 36 hours. A pioneer in the field is Maine Coast Sea Vegetables (established in 1971) in Franklin, Maine. They buy from independent harvesters, harvest from an aquaculture site in Frenchman Bay, and market wild and farmed sea vegetables worldwide.

After the collapse of the herring industry in the 1990s, and the closing of the sardine packing plants, fishermen kept on lobstering and shellfishing, and took to aquaculture and fish farming, and exporting dulse and other sea vegetables. As the market for seaweed grows, there is worry about overharvesting, which could upset the ecosystem of plants, seabirds, and marine creatures, and increased soil erosion along the shore. Maine's Seaweed Council, Sea Grant, and the Department of Resource Management work with seaweed harvesters to promote sustainable harvesting.

Meanwhile, at winter farmers' markets like the one in Saco, you'll find people selling seaweed — everything from snacks to a granola bar made with blueberries, seaweed, and chocolate. Many small, independent farmers are doing what they love — going out on small boats, respecting the clear, cold waters of Maine, living off the land and green acres at sea. As one farmer said, "farm living is the life for ME."

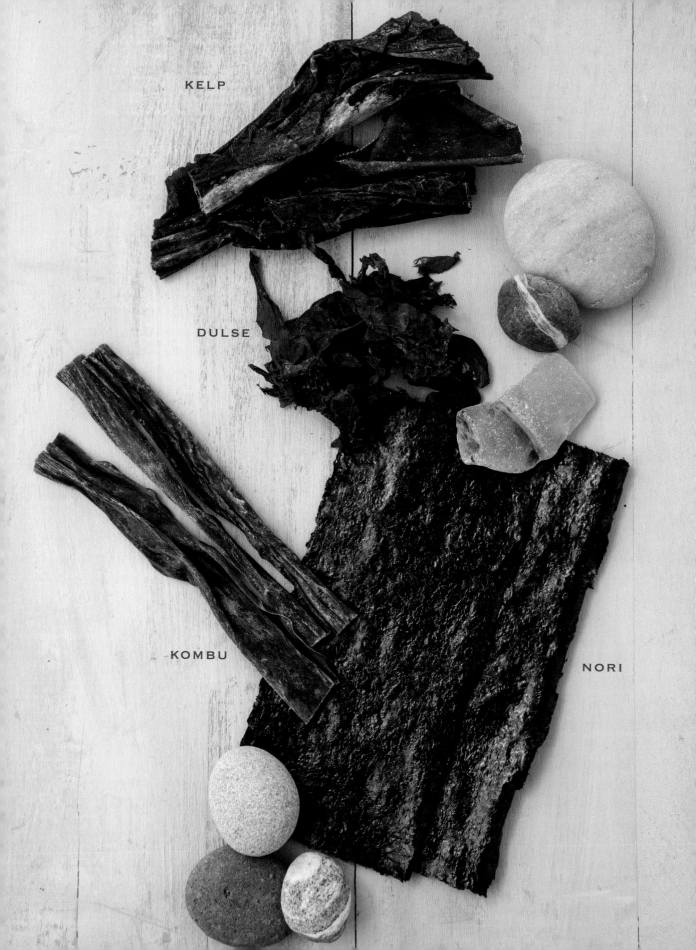

KELP

DULSE

KOMBU

NORI

COCKTAILS

Once the day is done — be it beachcombing, clamming, dog walking, sailing, or scrambling over rocks — it feels mighty good to sit back, enjoy the view, and have a cocktail. As the sun goes over the yardarm, you may want an easy drink or a carefully constructed cocktail. This chapter includes both. Serve simple appetizers that play with texture and temperature — a cool, creamy guacamole and chips go well with the grilled littlenecks (page 156); raw oysters on the half shell (page 150) might be served alongside smoked bluefish dip with crackers (page 88). With a grill nearby or a beach fire pit handy, cooking seafood appetizers is an impressive feat that is surprisingly easy. Fill in with little nibbles — bowls of wasabi peas, olives, nuts, and gherkins. When we're near the water, we want to enjoy the view, and that means simplifying.

What makes a great cocktail — a drink that lights up your guests' eyes and makes you happy to be delivering it? One that suggests promise and a wonderful evening ahead? Here are a few key components:

- Strive for balance. No one likes a drink that knocks them off their feet after four sips. There's a marvelous ditty in the Caribbean about rum punches, and how they should include sour, sweet, strong, and weak. Think of those components in making a drink.

- Think of your ingredients. Buy the best liquor you can afford. If using herbs (a sprig of basil, a stem of lavender) pick them fresh from the garden. Make sure the tonic and seltzer are fresh. Have enough ice.

- Go easy on the citrus. But remember to garnish.

- Glassware matters. If you're into the Ball jar look, go for it. But you can also scour antique shops and flea markets for amazing finds. Some of my favorites: shark highball glasses from the '50s, gold monogrammed champagne glasses from the '40s, rooster glasses from the '60s.

At the end of the day, a good cocktail can tell a story through the ingredients, the mixology, the name, the embellishments. Think of the design of this thing you've created — the glass, the ice, the garnish, the color. Wow your guests with the presentation. It's a great start to an evening.

Jalapeño-Cucumber Margarita

Making a margarita muddled with fresh jalapeños and cilantro raises the bar, but you can truly impress your friends by infusing your own tequila with jalapeños or habaneros. Just stem, de-seed, and quarter a few fresh chiles, place them in a liter bottle of 100 percent agave blanco tequila, and reseal. Let it steep for 6 to 24 hours: the longer it steeps, the stronger it will taste (you can test it along the way). Strain the tequila through a cheesecloth and store in a cool, dry place for up to a month.

½	cup tequila blanco
1	cucumber, sliced into ¼-inch rounds
½	jalapeño, halved and seeded, plus 4 jalapeño rounds for garnish
	Small bunch cilantro
½	cup lime juice
¼	cup Cointreau or Grand Marnier
	Lime wedge
	Citrus salt (orange and lemon peel and salt)

1. Combine the tequila, half the cucumber, the jalapeño, and the cilantro in a shaker and muddle until the cucumber and jalapeño start to break up. Add the lime juice and Cointreau and fill the shaker with ice.

2. Shake the mixture vigorously for 10 seconds, then strain into ice-filled margarita glasses that have been rubbed around the edge with lime and dipped in the citrus salt. Garnish with rounds of cucumber and jalapeño on the rim and in the glasses.

SERVES 4

Hurricane

When I was growing up on the WeWeantic River, hurricanes were an unusual and exciting occurrence. We'd board up the screen windows with plywood, haul the boats, and head for higher ground. Across the river was an upside-down house from the hurricane of '38. All that was left was the frame, which was resting on the roof rafters, rotting on the river marsh; it was a continuing source of curiosity and intrigue to neighborhood kids. We called it The Upside Down House.

Although Governor William Bradford of Plymouth Colony wrote in 1635 of a great tidal wave and mighty storm that swept through Buzzards Bay and was fearful to behold, the next devastating storm wasn't recorded again until 1815, when a "great gale" swept through the region, "prostrating" houses and destroying saltworks and ships. As late as the 1900s, hurricanes were so rare that New Englanders thought they only occurred in the tropics. Many had no idea what a hurricane was, which was what made the hurricane of '38 so devastating in Connecticut and Rhode Island.

Now, it's the rare autumn when we don't feel the threat of a hurricane. After living through one, a strong drink is in order. If the weather really deteriorates, revert to a Dark and Stormy (page 317).

1	ounce dark rum
1½	ounces light rum
1½	ounces passionfruit syrup
½	ounce lime juice
2	teaspoons superfine sugar
	Splash of grenadine
	Orange and/or lime wedge
	Cherry

Combine the dark rum, light rum, syrup, lime juice, sugar, and grenadine in a shaker over ice. Shake and strain into a highball glass filled with ice. Garnish with an orange or lime wedge and a cherry, and serve immediately.

SERVES 1

'38

Without warning, on the autumnal equinox of 1938, a Category 3 hurricane bore down on the New England coast, the worst of it pounding Rhode Island, eastern Connecticut, and southern Massachusetts.

The storm started on September 4 as a tropical disturbance near the Cape Verde islands. It spent almost two weeks gathering muster across the Atlantic, and on September 16, the captain of a Brazilian freighter warned the U.S. Weather Bureau of the storm he encountered northeast of Puerto Rico. Forecasters warned Floridians, who battened down the hatches.

But it didn't hit land; the storm veered north, traveling parallel to the Eastern seaboard. Charlie Pierce, a rookie forecaster in Washington, D.C., at the U.S. Weather Bureau, was sure the storm would hit New England and wanted to warn people, but his supervisor overruled him; they had already been embarrassed by putting Florida on full warning with no storm materializing.

No regional warnings were broadcast. In 1938, many New Englanders didn't even know what a hurricane meant — the tradition of naming storms didn't begin until 1950 — or even how to pronounce it. There had been a storm in 1635 (which the Pilgrims wrote was apocalyptic) and the Great September Gale of 1815, but many didn't believe a tropical cyclone could hit the cold waters of the North Atlantic.

With Europe on the brink of war, people on September 21 were listening to Franklin Delano Roosevelt's radio address, which stations were loath to interrupt. There were no weather balloons, weather satellites, or radio buoys to forecast the hurricane's approach. Around 2:30 P.M., the hurricane made landfall on the south shore of Long Island at high tide.

One woman who lived across the road from the beach was hosting her daughter's birthday party, and watched in disbelief as a wall of water came toward her house. She scooped the kids up to the second floor, and then to the roof. Waves were 40 feet tall, swallowing up whole communities as the hurricane hit Connecticut.

It gained intensity passing into Rhode Island, with winds over 120 mph and a 15-foot storm surge in Narragansett Bay. Providence was 13 feet under water. By the time the monster storm reached Massachusetts, weathermen were clocking gusts of 186 mph.

Five hundred and sixty-four people were killed, another 1,700 were injured, nearly 9,000 buildings were destroyed and another 15,000 damaged. The force was equivalent to an H-bomb going off every minute, according to some scientists. Winds were so strong that chickens were plucked of their feathers, boats were tossed like toothpicks, and cottages ended up blocks away. (Interestingly, Cape Cod was spared, the canal providing an outlet for the high tide, and the high clay cliffs on the southern side of the Cape forming a protective barrier.)

Old-timers remember the hurricane of '38, and even those of us who weren't around have a healthy respect for the region's biggest weather event of the 20th century.

Boat Drink

Jimmy Buffett coined the term "boat drinks" — those colorful, rum-saturated umbrella drinks that slide us to St. Somewhere with a palm-fringed cove. Sometimes called tiki drinks, they usually feature rum and at least one fruit juice, and they hit the spot after a day on the water.

1½	ounces pineapple juice
1½	ounces orange juice
1	ounce white rum
1	ounce golden rum
1	teaspoon grenadine
	Nutmeg (optional)
	Wedge of lime

Fill a tall glass with ice and pour in the pineapple juice, orange juice, white rum, golden rum, and grenadine. Stir gently. Grate a bit of nutmeg on top if you have it, and squeeze the lime wedge into the drink, then drop in the wedge. Serve immediately.

SERVES 1

THE WEEKAPAUG

An elegant seaside inn in Westerly, Rhode Island, may be the only New England hotel with a naturalist on staff. Mark Bullinger sat down with me at breakfast and told me why white snowy owls were appearing along the eastern seaboard that spring and why the caverns off Block Island make for great swordfishing. Each guest room has a Sibley's bird book and binoculars, and in the lounges there are hundreds of books about the sea, the region, and boats. The naturalist said his first job upon being hired was to spend $3,000 on books. It shows. You can also browse through old inn guest books dating back to 1899, where every guest (and often the chauffeur) would sign in, including Franklin Delano Roosevelt one summer.

Weekapaug Bee Keeper

The bartender at the Weekapaug Inn has a sure hand, as evidenced by this drink. The honey-infused gin is from Vermont. Rim the glass with candied ginger and sugar for the full effect. Feel free to adjust the ratio if you prefer more or less honey.

3	mint leaves plus 1 for garnish
1¾	ounces Barr Hill Gin
½	ounce lime juice
½	ounce local honey
	Ginger sugar (granulated sugar blended in the food processor with crystallized ginger)

1. Muddle the 3 mint leaves in a chilled cocktail glass.

2. Combine the gin, lime juice, and honey in a shaker with ice and shake vigorously. Strain into the cocktail glass and garnish with a mint leaf.

SERVES 1

RUM-RUNNING IN WESTPORT

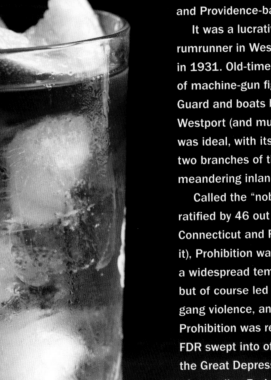

During Prohibition (1919–1933) when you weren't allowed to make, sell, or possess liquor, a new business flourished along the eastern seaboard: rum-running, where local fishermen and boat owners hauled booze from ships that had come in from Canada and anchored in "Rum Row" (an area just outside U.S. jurisdiction in territorial waters). Small boats would load up, then race to shore trying to avoid the Coast Guard.

"The line of schooners around Block Island looked like a city," reminisced one Westport (Massachusetts) rumrunner to the local paper in the 1970s. Booze was hidden in farmhouses, barns, and garages until it could be distributed by New York– and Providence-based gangsters.

It was a lucrative business; one rumrunner in Westport made $15,000 in 1931. Old-timers still tell stories of machine-gun fights with the Coast Guard and boats being blown up. Westport (and much of the waterfront) was ideal, with its craggy shore and two branches of the Westport River meandering inland.

Called the "noble experiment" and ratified by 46 out of 48 states (only Connecticut and Rhode Island fought it), Prohibition was the consequence of a widespread temperance movement, but of course led to the rise of crime, gang violence, and other problems. Prohibition was repealed in 1933, after FDR swept into office in 1932 during the Great Depression on the platform of repealing Prohibition.

"Later my financial position improved, and I bought rum, which solved most of my drinking problems. It also made my cooking taste better."

— BRITISH ADVENTURER
FRANCIS BRENTON

Dark ᴬᴺᴰ Stormy

In the mid 1800s, with the British Royal Navy dockyard in Bermuda, it was standard practice for the navy to allocate rum rations to soldiers. About the same time, the British Gosling Brothers began marketing in their Bermuda rum distillery a heavy blend of dark rums that eventually became Gosling's Black Seal rum. Whalers from Nantucket tasted the rum, and brought the recipe back home. The name is said to have come from a sailor who, holding up the drink, noted that it was "the color of a cloud only a fool or a dead man would sail under." Traditionally made with Gosling's Black Seal Rum and ginger beer, this cocktail is also delicious with Hurricane Rum, a spirit made on Nantucket in small batches by Triple Eight Distillery during hurricane season (that would be late summer–early fall). Newport yachtsmen still prefer Dark and Stormies after a run to Block Island. Any port in a storm!

    6   ounces ginger beer (Reed's, Maine Root, or Gosling's work well)
    1½   ounces Triple Eight Distillery Hurricane Rum or Gosling's Black Seal Rum
          Lime wedge

Pour the ginger beer into a tall glass with ice. Float the rum on top and squeeze the lime into the glass. Stir with a swizzle stick and serve.

SERVES 1

Made in ᵀᴴᴱ Shade

It's late August and your seaside garden is voluptuous with fragrant basil. When it rains, you pad out barefoot, just to smell it. When it's time to harvest, you make pesto, of course, and eat tomato salads every night, but what else? A basil gin cocktail. Which you can serve at your pesto-making/eating party.

    10   basil leaves
    4   ounces Hendrick's gin
    2   ounces lime or lemon juice
    2   ounces simple syrup or Lavender Syrup (page 322)

Reserve two of the prettiest basil leaves. Tear the other eight leaves, and then with a wooden spoon, muddle them in a shaker until they are bruised and fragrant. Add the gin, lime juice, syrup, and some ice, and shake vigorously until the outside of the cocktail shaker is freezing cold. Strain into two short ice-filled glasses and garnish each with a basil leaf.

SERVES 2

Tew's Mojito

In 2007, four guys started brewing spirits at the Newport Distillery, and their Thomas Tew Rum is the first to be made in Rhode Island in 135 years. They picked the name from a historical character they found in the library, which I thought was rather amusing since, according to our family tree, my great-great-great-great-grandfather William Tew (1745–1808) was captain of the Continental Army for George Washington (he carried the Rhode Island payroll to Valley Forge). I bet the army wished he also brought rum.

6	mint leaves
1½	ounces golden or dark rum
	Juice of ½ lime plus a wedge for garnish
2	tablespoons Lavender Syrup (page 322) or sugar
1–2	ounces club soda

Muddle the mint leaves in a rocks glass to release the oils. Leave the mint in the glass and add ice almost to the top. Pour in the rum, lime juice, and lavender syrup, and stir. Top with the club soda, stir, and taste, adding more sweetener if desired.

SERVES 1

RUM

The history of rum in New England runs deep. Invented in the West Indies in the 1600s, rum caught on in the Colonies, and distillers began importing molasses and creating a major industry. By the time the Molasses Act of 1733 was passed, there were 159 rum distilleries in New England, and authorities estimated that American colonists each drank about 3.75 gallons of rum annually. In 1769, there were 22 distilleries in Newport alone. By 1842, however, as whiskey took over as the spirit of choice (corn was more reliable than molasses after the abolition of the slave trade in 1808), there were none.

The New Englander

This bracing, astringent drink is as clipped as a Yankee's accent, as sharp as a bowsprit, and as soothing as the dunes under a full moon on a summer night. Over the years, it's been called a Bog Fog, a Cape Codder, even a New Yorker. If you're making it with 100 percent cranberry juice, you might want to add a bit of sugar or simple syrup.

5	ounces cranberry juice
2	ounces top-shelf vodka
	Dash of Rose's lime juice or a squeeze of half a lime
	Wedge of lime

Pour the cranberry juice, vodka, and lime juice into a tall glass with ice. Swirl and serve, garnished with a wedge of lime.

SERVES 1

319

CITRUS SALT

It's so easy to make, and what a tasty gift. First zest your favorite citrus, or a combination — lemon and orange, lemon and lime, grapefruit, etc. Put the zest in a bowl with some salt (you can use any kind of salt — fine, flaky, coarse, kosher). The less salt, the stronger the citrus flavor, so play with the amounts. (You might start with one-third zest to two-thirds salt.) Using your fingers, break up any clumps and rub the citrus into the salt to distribute the oils. Spread the mixture in a thin layer on a tray and leave at room temperature overnight to let the salt dry the zest. Kept in a sealed container, citrus salt should last 2 weeks.

Citrus salt works well whenever you want a bright citrus finish — not only to rim drinks, but to toss into soups, sprinkle on hard-boiled eggs, or lace into cookies or caramel sauce. I know someone who even puts a pinch on ice cream.

Strawberry Bellini

This drink is particularly tasty in June, when fresh, tender strawberries come in. If you've picked your own strawberries, you know that they are incredibly short-lived, and after you've made shortcake, and put them in your cereal, and maybe made a pie, and eaten half of them warm from the vine, you probably still have some left over. This marvelous drink is for those times. There also seem to be so many good excuses to celebrate in June — graduations, recitals, birthdays, weddings, baby showers.

The original bellini, which was made with prosecco and peach nectar, was invented at Harry's Bar in Venice in the 1930s. He considered it a seasonal specialty, although it became so popular that these days, it's served year-round.

1	cup fresh strawberries plus 2 ripe small strawberries, cleaned and hulled
½	teaspoon fresh lemon juice
1	teaspoon sugar
	Prosecco or champagne, chilled

Blend and purée the cup of strawberries with the lemon juice and sugar. Pour into two glasses, filling them halfway, and fill the rest with champagne. Garnish each glass with a strawberry.

SERVES 2

SIMPLICITY REIGNS

The beach is the place to let go, to walk barefoot and enjoy the view and the company. Having a few drinks with friends? Set out simple snacks that satisfy the craving for salt. Thick-cut potato chips, olives stuffed with anchovies, carrots or broccoli with hummus. It needn't be fancy. I learned from living on boats that it's the view and the company that matter.

Lavender Gin *with* Bubbles

I learned from the Weekapaug bartender that, when serving a cocktail topped off with soda, carbonation is key, for it releases the drink's aroma right below our nose. Moreover, different carbonation levels yield different-sized bubbles (not to mention minerality). Who knew? Club soda or seltzer give you larger, stronger bubbles, whereas San Pellegrino is more delicate. Good bartenders know this and play with the subtle distinctions in carbonation — what do you bet there comes the day when folks at the bar are as intrigued by their bubbles as they are by their bitters and gin? "I'll take top-shelf carbonation, please!"

	Wedge of lemon plus small half-circle
2	ounces dry gin
1–2	tablespoons Lavender Syrup (recipe follows)
	Bubbly water of choice
	Sprig of lavender (optional)

Squeeze the lemon juice into a highball glass full of ice and drop the wedge into the glass. Add the gin, lavender syrup, and bubbly water. Stir, then top with the half circle of lemon, and a sprig of lavender if you have one, and serve.

SERVES 1

Lavender Syrup

If you have a coastal garden, you know which plants are rugged and tolerate sea spray, storms, and winds. Lavender thrives by the sea, and this is a simple syrup that makes good use of lavender. I make a big batch and keep it in the refrigerator as a sweetener for many beverages — my son, Trainer, especially loves it in his tea.

1	cup water
1	cup sugar
3	tablespoons dried lavender buds

Combine the water, sugar, and lavender in a saucepan and bring to boil, then turn off the heat and let the lavender steep in the liquid for 30 minutes or so. Strain and transfer to a glass jar. This simple syrup will keep for several weeks refrigerated.

MAKES 1⅓ CUPS

Pink Gin

Made by distilling herbs, roots, and other ingredients, bitters add a dash of flavor and enhance other ingredients. True to its name, the botanical flavor is slightly melancholy, even bitter. The granddaddy of them is Angostura, first made in Venezuela in the 1800s as a medicinal tonic, and it's long been the favored drink of British Navy seamen, thought to ward off seasickness and hangovers. There's a bitters boom these days, from Coastal Root Bitters in Portland to the Black Trumpet Bistro in Portsmouth, New Hampshire. Rhubarb bitters? Yep. Cranberry bitters? Of course.

1½	ounces gin
2–3	dashes of Bittermens Boston Bittahs or Angostura bitters
	Strip of lemon peel

Pour the gin and bitters into a cocktail shaker with ice cubes and shake briefly. Strain into a chilled cocktail glass filled with ice and garnish with lemon peel.

SERVES 1

7

SWEET

ENDINGS

THE IRONY HERE IS that our family didn't have desserts growing up. Oh, sure, my grandmother's brownies were famous at family parties and beach picnics, but my *mother* never baked. To her, baking was just shy of mending, half a notch above cleaning, dangerously close to ironing, and she steered clear of anything that might pigeon-hole her as a housewife in the 1950s. If company came, she'd buy "squashed-fly" (oatmeal raisin) cookies for dessert. When she stapled the hem of one of my dresses, my father took over, bringing out his sailcloth needles to sew my Girl Scout badges.

Thankfully, my grandmother moved in with us when I was in ninth grade. She loved to bake, and I followed in her footsteps in that regard. (Although, unlike many in our family, she didn't have a nautical bone in her body. She mistakenly referred to my father's Herreshoff Beverly Dinghy as his Beverly Dinghus, and he hooted with laughter when — after overhearing us discussing an upcoming sailing trip and making the tricky passage through Quick's Hole to the Vineyard — she later asked how our trip to Martha's Hole turned out.)

This chapter includes desserts, but it's really a broader brush of sweet New England specialties. Take Cranberry Granola, for example, which is delicious for breakfast but also doubles as an ice-cream topping, not to mention hiking snack. In some way, New England's changing seasons ("Don't like the weather? Just wait a minute.") can be told through a calendar of the harvest — picking strawberries, which leads to low-bush blueberries, which is followed by peaches, which culminates in cranberries. This chapter highlights the seasonal succession of New England fruits as inspiration for traditional and non-traditional desserts and condiments.

Drenched Blueberry Cake

In New England, rhubarb marks the beginning of the growing season, while cranberries bookmark the end. Blueberries settle in somewhere in between. There are plenty of places to pick wild blueberries, and they shine in this cake.

CAKE

2	cups all-purpose flour
1	cup sugar
2½	teaspoons baking powder
3	tablespoons butter, melted
½	teaspoon vanilla extract
⅔	cup milk
1	egg
2	cups fresh blueberries

SAUCE

½	cup unsalted butter
1	cup sugar
¾	cup heavy cream
½	vanilla bean, split lengthwise

1. Preheat the oven to 350°F (180°C) and butter a 9-inch round pan.

2. To make the cake: Sift the flour, sugar, and baking powder into a large bowl, and then add the butter, vanilla extract, milk, and egg. Beat for 2 minutes, or until smooth, and then stir in the fruit. Pour into the prepared pan, and bake for 1 hour, or until a toothpick comes out clean.

3. Transfer to a wire rack to cool, then remove to a plate.

4. To make the sauce: Melt the butter in a small saucepan over low heat. Add the sugar, cream, and vanilla bean, stirring to mix well. Cook, stirring occasionally, for 5 minutes. Remove the vanilla bean.

5. Slice the cake and serve with individual portions generously drenched in warm sauce.

SERVES 8

Blueberry Peach Cobbler *with* Vanilla Bean

In early September, the days are still warm, but fall is in the air. Some berries and plenty of peaches abound, and the fall fruits — apples, grapes, pears — are coming in. It's a time for biking and eating in the countryside, maybe even overeating, since you know the first fall freeze will end it abruptly.

FILLING

2	cups blueberries
6–7	ripe peaches with skin, cut into segments
2	tablespoons tapioca
⅔	cup granulated sugar
3	tablespoons brown sugar
	Juice of 1 lemon
	Seeds scraped from 1 vanilla bean
⅛	teaspoon nutmeg

BISCUIT DOUGH

2	cups all-purpose flour
2	tablespoons granulated sugar
2	teaspoons baking powder
¼	teaspoon salt
6	tablespoons cold unsalted butter, cut into cubes
1	cup buttermilk

TOPPING

1	tablespoon granulated sugar
½	teaspoon cinnamon

1. Preheat the oven to 350°F (180°C). Lightly butter a 9- by 13-inch baking dish.

2. To make the filling: In a mixing bowl, combine the blueberries, peaches, tapioca, both sugars, lemon juice, vanilla bean seeds, and nutmeg, and stir to coat the fruit.

3. To make the biscuit dough: In a separate bowl, combine the flour, granulated sugar, baking powder, and salt. Using your fingers, add the butter and mix into the dry ingredients until you have a coarse crumble. Add the buttermilk and stir until the biscuit dough comes together.

4. To make the topping: Combine the granulated sugar and cinnamon in a small bowl.

5. Pour the fruit filling into the prepared baking dish, spreading it out to fill the dish and make a level surface. Place the biscuit dough in ½-inch-thick clusters on top of the filling. Dust with the topping. Bake for 50 to 60 minutes, until the biscuit is cooked and golden. Serve warm.

SERVES 8

Blueberry-Ginger Tart

This is one of my all-time favorite desserts. A version of this recipe originally appeared in *The Yachting Cookbook*, a book I wrote with Liz Wheeler after we had many adventures delivering Hinckley sailboats together from Southwest Harbor, Maine, to the British Virgin Islands.

PASTRY

1½	cups all-purpose flour
3	tablespoons firmly packed brown sugar
1	teaspoon ground ginger
¼	teaspoon salt
½	cup (1 stick) butter, softened

FILLING

½	cup candied ginger
4	tablespoons granulated sugar
3	tablespoons all-purpose flour
1	envelope (¼ ounce) unflavored gelatin
¼	teaspoon salt
4	egg yolks
1½	cups milk
1	teaspoon vanilla
½	cup heavy cream
1	pint blueberries
¼	cup peach or strawberry jam

1. Preheat the oven to 375°F (190°C).

2. To make the pastry: Combine the flour, brown sugar, ginger, and salt in a bowl and cut in the butter until the mixture resembles coarse crumbles. Knead the mixture by hand until blended. Pat the pastry along the bottom and sides of an ungreased 10-inch tart pan with a removable bottom. Bake 20 minutes. Transfer to a wire rack to cool (keeping the sides of pan on).

3. To make the filling: Pulse the candied ginger with 1 tablespoon of the granulated sugar in a food processor until the ginger is finely ground. Set aside. In a saucepan, combine the remaining 3 tablespoons granulated sugar, flour, gelatin, and salt. In a bowl, beat the egg yolks with the milk until well blended, and then stir into the sugar mixture. Cook over medium-low heat, stirring constantly, until the gelatin dissolves and the mixture thickens, about 15 minutes (do not boil). Remove the saucepan from the heat and stir in the candied ginger and vanilla. Refrigerate for an hour, or until mixture mounds slightly.

4. In a small bowl at medium speed, beat the heavy cream until soft peaks form. Fold the cream into the custard filling and spoon into the pastry shell. In a small saucepan warm the jam until melted. Cover the custard with the blueberries and brush with the jam. Refrigerate the tart until the custard is set, about 1 hour. Keep stored in the refrigerator.

SERVES 10

"If a man will go at Christmas to gather cherries in Kent he may be deceived, though there be plenty in summer; so, here these plenties have each their seasons."

— CAPTAIN JOHN SMITH, 1616, DESCRIBING THE COD, FRUIT, AND HERBS HE FOUND FROM PENOBSCOT BAY TO CAPE COD

Apple Pie

The tradition of baking pies was brought to America by the Puritans, and it continues to this day — it's hard to imagine a Thanksgiving dinner without at least one. Early colonists used English-style ingredients, such as apples, which they brought to America in the 1600s. (The first orchard was planted in Boston in 1621; only crabapples are native to New England.) Soon, seeds of various varietals spread along Indian trade routes and were cultivated in colonial orchards. Interestingly, a list of provisions that settlers to New England were advised to take aboard ship in 1630 included sugar, cloves, cinnamon, mace, and nutmeg — all tropical in origin. These spices are to this day customary flavors in Thanksgiving pies.

PASTRY

2	cups all-purpose flour
½	teaspoon salt
1	cup (2 sticks) unsalted butter, chilled and cut into pieces
4–6	tablespoons cold water

FILLING

⅔	cup sugar
3	tablespoons all-purpose flour
1	teaspoon cinnamon
½	teaspoon nutmeg
1	teaspoon vanilla extract
⅛	teaspoon salt
5	medium tart apples, peeled, cored, and quartered
3	tablespoons butter, cut into 4 pieces
1	egg white

1. To make the pastry: In a food processor, mix the flour and salt. Add the butter and pulse until the mixture resembles small peas. Sprinkle with cold water, a tablespoon at a time, until the dough starts to stick together. Gather the pastry into a ball, divide in half, and flatten into two rounds on a floured surface. Wrap in plastic wrap and refrigerate until the dough is firm, about 30 minutes.

2. Preheat the oven to 425°F (220°C). With a floured rolling pin, roll out the two pastry discs, making one slightly larger than the pie plate. Ease the larger crust into the pie plate, pressing against the bottom and sides.

3. To make the filling: In a bowl, combine the sugar, flour, cinnamon, nutmeg, vanilla, and salt. Fold in the apples and toss to coat. Pile the filling into the bottom crust and dot the butter on top in four quadrants. (Set the bowl aside without rinsing.) Cover the filling with the other crust, pinch the edges, and cut several steam vents in the top.

4. Toss the egg white into the bowl (which should still have some sugar and spices in it) and with your fingers swirl the white with the spices, then brush onto the top crust. (Save the yolk for another use, or if you have chickens, you can feed the egg yolk and the crushed shells to your hens.)

5. Bake for 10 minutes, then reduce the temperature to 350°F (180°C). Continue baking until the crust is brown and the filling is bubbling, about 45 minutes.

SERVES 6

Thumbprint Cookies with Beach Plum Jam

Beach plums were one of the first plants the colonists saw when they came ashore in the 1600s. Native to the sandy North Atlantic coast, they can be found from Newfoundland to North Carolina, and are most prevalent from Massachusetts to New Jersey. They're great for erosion control and as a pretty ornamental, and the berries, which ripen in August, are suitable for making jam. From Maine to Connecticut, they grow wild — along roads, in sandy parking lots, backyards, and on the back sides of dunes. For many, foraging for beach plums is just another wonderful day at the beach. At roadside stands and in gift shops, you can find beach plum jam put up locally, with homemade labels. There's even a Jam and Jelly Shop in Chatham, Massachusetts. You can substitute rose hip jam or other fruit jams for beach plum in these cookies if need be.

⅔	cup unsalted butter
⅓	cup sugar
3	egg yolks
1	teaspoon vanilla extract
½	teaspoon salt
1½	cups all-purpose flour
2	lightly beaten egg whites
¾	cup toasted chopped pecans
⅓	cup beach plum jam

1. Cream the butter and sugar until fluffy. Add the egg yolks, vanilla, and salt. Beat well, add the flour, and mix well.

2. Preheat the oven to 350°F (180°C).

3. Using a tablespoon measure, shape the dough into balls, then dip into the egg whites, and roll in the nuts. Place 1 inch apart on an ungreased baking sheet and press down the center with your thumb. Bake 15 to 17 minutes, or until golden. Let the cookies cool, then fill the centers with jam.

MAKES 30 COOKIES

Cranberry Granola

Cranberry granola is a specialty at the Black Dog, a merchandising juggernaut on Martha's Vineyard that started as a loafers-and-fringed-jeans kind of wharfside restaurant in Vineyard Haven in 1971, back when there were no year-round restaurants on the island. The owner, a former Air Force pilot, had retired on the island, restored a 19th-century cutter (the *Shenandoah*), and vowed to get a dog that would sail with him, no matter the weather. The company's black lab–boxer mascot was that beloved dog.

This recipe features cranberries and maple syrup, two specialties from New England. Besides being breakfast, this granola can be enjoyed as a topping for vanilla ice cream or fruit salad.

3	cups old-fashioned rolled oats
1	cup combined unsalted nuts (almonds, cashews)
1	cup unsalted seeds (sunflower, sesame, flaxseed, etc.)
1	tablespoon wheat germ
½	teaspoon salt
4	tablespoons unsalted butter
½	cup maple syrup
1	teaspoon vanilla extract
⅓	cup firmly packed brown sugar
1	cup dried cranberries or blueberries

1. Preheat the oven to 350°F (180°C).

2. In a large bowl, combine the oats, nuts, seeds, wheat germ, and salt.

3. In a small saucepan, combine the butter, maple syrup, vanilla, and brown sugar. Cook over medium heat until the butter is melted. Pour over the dry mixture and stir until blended. Spread the mixture out flat on a baking sheet. Bake for 20 minutes, stirring halfway through.

4. Let the granola cool and then stir in the cranberries. Stored in a sealed container, it will keep for a month or more.

MAKES 6 CUPS

Cranberry Applesauce

Applesauce is not just for kids; it's delicious with roasted meats or grilled fish, as a snack, or dessert. It's also so incredibly easy to make that you might wonder, after you compliment yourself on having made a batch, why people don't do it more often.

5	Macintosh apples, peeled, cored, and cut into eighths
½	cup fresh or frozen cranberries
2	tablespoons confectioner's sugar
½	cup water
	Pinch of salt
¼	teaspoon cinnamon

1. Combine the apples, cranberries, sugar, water, salt, and cinnamon in a saucepan and bring to a boil over medium-high heat. Reduce the heat to low, cover, and simmer, stirring occasionally, until apples are tender, 20 to 25 minutes.

2. Transfer the sauce to a food processor and pulse to desired consistency. Serve warm, at room temperature, or cold.

SERVES 4-6

336

CRANBERRIES

One of three fruits native to North America, cranberries grow wild on vines in sandy bogs and marshes along the shore. Eastern Indians called them *sassamanesh*, while Cape Cod Pequots and South Jersey tribes called them *ibimi*, or "bitter berry." Early German and Dutch settlers named them "crane berries" due to the flower's resemblance to the head and bill of a crane.

Sandy soil and natural peat holes that turn into bogs create ideal growing conditions for the cranberry vines. Relatives of rhododendrons, cranberry vines blossom in spring and bud in fall. The blossoms start off a beautiful pink hue, then a week later open up, turning white and creating what looks like a snowstorm on the bog (usually mid-June to mid-July). The chemical reaction that turns the color of leaves (in the autumn) also causes cranberries to turn from green to dark red and the vines to go dormant, enabling farmers to cull the berries.

Most of the older cranberry bogs are odd-shaped, made from maple and cedar swamps that were turned into bogs by cutting down the trees, putting down sand, and planting cranberries. Newer bogs are rectangular in shape; they are easier to manage and control.

Southeastern Massachusetts is bog country, and Ocean Spray is headquartered there, a cooperative formed in 1930 by three cranberry growers that today has more than 700 members nationwide.

Cranberry-Orange Relish

When I was a child we always had Thanksgiving at my cousin Jeff's house in Scituate, where we walked with the dog on the beach after dinner, collecting shells. Later, the intergenerational feast moved to Mattapoisett, where my parents hosted whatever cousins, aunts, and uncles could come. The tradition had a lasting effect. Though now grown, my cousins and I still sail and swim together in the summers when we can, and every fall we have a Thanksgiving feast at my cottage. The hardiest of us don waders to go quahogging one last time for chowder that simmers on the stove while we prepare the meal. After dinner, the adults sit around the fire, appreciating the last seaside hurrah before buttoning up our cottage for the winter, while the children play games on the lawn and fish off the point.

1	(12-ounce) bag fresh cranberries
1½	navel oranges, peeled and quartered
¼	cup slivered almonds
¼	cup sugar

Combine the cranberries, oranges, almonds, and sugar in a food processor and pulse until the desired consistency is reached. Letting the relish stand for a few hours at room temperature allows the flavors to blend more, although if you are in a hurry it can be served immediately.

MAKES 3½ CUPS

The Mud Hole

Pushing east from Mount Desert Island, there's a little cove 25 miles east of Schoodic Point called the Mud Hole. The sailing guidebooks have a lot to say about entering Mud Hole — the entrance closes off at low tide, but there's still 16 feet of water in the hole, making it a safe anchorage and a beautifully remote spot on the eastern shore of Great Wass Island, home to a Nature Conservancy preserve with rare plants, trees, seabirds, and songbirds. Chocolate soufflés have a temperamental reputation, but they are easy as pie . . . er, soufflés. Don't peek while cooking, and eat 'em fast — every lovin' spoonful.

2	tablespoons butter plus more for coating the baking dish
8	ounces semisweet chocolate, chopped
4	eggs, room temperature, separated
1	teaspoon vanilla extract
	Pinch of salt
¼	teaspoon cream of tartar
⅓	cup sugar plus more for baking dish
	Whipped cream, for serving

1. Preheat the oven to 350°F (180°C). Butter a 1½-quart souffle dish, then coat it with sugar, tapping out the excess.

2. In a double boiler set over simmering water, melt the butter and chocolate, then whisk in the egg yolks, vanilla, and salt. Remove from the heat.

3. Beat the egg whites until foamy, and then add the cream of tartar, continuing to beat until soft peaks form. Add the sugar slowly, beating until you have stiff, glossy peaks, about 5 minutes.

4. With a rubber spatula, gently and slowly fold the egg whites into the chocolate mixture. Scoop the soufflé into the prepared baking dish, smoothing over the top. Bake until puffed and set, about 30 minutes. Serve immediately with a dollop of whipped cream.

SERVES 6-8

ACKNOWLEDGMENTS

A PERSONAL BOOK like this is the culmination of many people, ideas, and experiences. Many thanks to those who contributed stories, reflections, and recipes to this book, including Jody Fijal, Donna Monaco Olsen, Nancy Thomas, Gay Gillespie, Estrella Montalvo, Marc DeRego, Mark Bullinger, Chris Johnson, and Rich Pasquill. I'm also grateful to Helen Bush Sitter for her beautiful coastal garden design and to Jane Duff Gleason for making a simple cottage shine. That we can see Jane's father's bench on the marsh is an added treat.

My heartfelt thanks, too, to Lisa Hiley, Margaret Sutherland, Carolyn Eckert, Catrine Kelty, Joe Keller, and the team at Storey Publishing, who understood the vision, plus are so much fun to go on location with and work on new projects. I'd also like to thank Heather Hobler, Ross Menard, Peter and Amy Trow, Jay Johnson, and Stephanie Boyd (whose pottery shines with the colors of the sea), not to mention Maur and Rog Brooks, whose friendship dates back to our lifeguarding days. Little did Maur expect, that summer day long ago when she asked if she should bring brownies and sourballs on a sailing picnic after work with my parents, that my father would retort, "Maureen, I take my balls wherever I go!" Nor could I imagine this book without the wonderful Aplin tribe, good friends with whom we've shared so many stories, milestones, islands, and coastlines. I'd break the bake with you any day.

This last paragraph is reserved for a special nod to Peg, Joe, and my coastal cousins — Jeff, Lynn, Sue, Steve, Doug, and Lee — with whom I share so many memories (glad you never clam up) and anticipate many more to come. My love and gratitude always to Isabel and Trainer, who are honest recipe testers and have had sand between their toes since they were babies. And, boy, of all the books I've written — from nuclear power, to the unified field theory, trails of flame, tiki drinks, and chickens — wouldn't my parents have loved this one. Here's to you, RHB and FOB.

"We shall not cease from exploration, and the end of all our exploring will be to arrive where we started and know the place for the first time."

— T. S. ELIOT

INDEX

Page numbers in *italic* indicate photos.

C

342

D

351

INDEX